6010

FOR REVIEW
August, 2013
S. EJAZ AHMED
BOOK REVIEW EDITOR
Technometrics

Radiation-Induced Processes of Adaptation

Victoria L. Korogodina • Boris V. Florko
Ludmila P. Osipova

Radiation-Induced Processes of Adaptation

Research by statistical modelling

Springer

Victoria L. Korogodina
Joint Institute for Nuclear Research
Moscow, Russia

Boris V. Florko
Joint Institute for Nuclear Research
Moscow, Russia

Ludmila P. Osipova
Institute of Cytology and Genetics SB RAS
Novosibirsk, Russia

ISBN 978-94-007-6629-7 ISBN 978-94-007-6630-3 (eBook)
DOI 10.1007/978-94-007-6630-3
Springer Dordrecht Heidelberg New York London

Library of Congress Control Number: 2013939262

© Springer Science+Business Media Dordrecht 2013
This work is subject to copyright. All rights are reserved by the Publisher, whether the whole or part of the material is concerned, specifically the rights of translation, reprinting, reuse of illustrations, recitation, broadcasting, reproduction on microfilms or in any other physical way, and transmission or information storage and retrieval, electronic adaptation, computer software, or by similar or dissimilar methodology now known or hereafter developed. Exempted from this legal reservation are brief excerpts in connection with reviews or scholarly analysis or material supplied specifically for the purpose of being entered and executed on a computer system, for exclusive use by the purchaser of the work. Duplication of this publication or parts thereof is permitted only under the provisions of the Copyright Law of the Publisher's location, in its current version, and permission for use must always be obtained from Springer. Permissions for use may be obtained through RightsLink at the Copyright Clearance Center. Violations are liable to prosecution under the respective Copyright Law.

The use of general descriptive names, registered names, trademarks, service marks, etc. in this publication does not imply, even in the absence of a specific statement, that such names are exempt from the relevant protective laws and regulations and therefore free for general use.

While the advice and information in this book are believed to be true and accurate at the date of publication, neither the authors nor the editors nor the publisher can accept any legal responsibility for any errors or omissions that may be made. The publisher makes no warranty, express or implied, with respect to the material contained herein.

Printed on acid-free paper

Springer is part of Springer Science+Business Media (www.springer.com)

Dedicated to Vladimir I. Korogodin

Foreword

Since the last century, man has made extensive use of nuclear energy. From time to time, radiation accidents caused by the malfunction of nuclear power plants have occurred. Radionuclides that originated with accidents in Chernobyl and Fukushima have been detected around the world. Radiobiology has always been topical; it has obtained special significance today. Investigations of the last two decades have revealed bystander effects and genomic instability induced by radiation stress that increased the effect of low doses.

In ecology, it is difficult to determine the influence of low radiation sources on organisms by standard biological methods: we can often simultaneously register reduction of the organisms' survival in the population and reduction in the frequency of their cells with chromosomal abnormalities. This paradox leads us to the question: "Is radiation stress harmful or useful?" An approach to understanding low-dose effects is the use of statistical modelling, which was the method used by the authors of this book.

Statistical modelling is widely used for solving physical problems. In biology, these methods were used by R.A. Fisher, and developed by H.A. Orr. It was shown that organisms are geometrically distributed according to their fitness in a given environment. In this book, the methodology of statistical modelling is used in radiobiology. The authors consider distributions of irradiated organisms and cells on the number of abnormalities. The observed regularities reflect adaptation processes coupled with instability. The values of distributions can be used for estimation of damaging and death processes of cells and organisms in populations that experienced radiation stress.

The strength of these investigations is in the experimental basis of the calculations. Consideration of the ideas of bystander effects and methods of statistical modelling has suggested that irradiation that does not exceed natural background is associated with adaptation and instability processes. These effects were demonstrated in the ecological samples of plant populations growing around the nuclear plant. The authors also showed that chromosomal instability continues in progeny of people living in areas that experienced fallout from nuclear tests 50–60 years ago. This method makes it possible to calculate risks of chromosomal instability more correctly than on average biological values.

To summarize, the developed methodology can be used not only for radiation ecology and epidemiology, but in radiotherapy, ecology, sociology and other fields for the consideration of processes induced by environmental factors.

Professor at McMaster University   Prof. Colin B. Seymour
Hamilton Ontario, Canada

Preface

Our investigations began in the final decade of the last century when radiobiologists were debating whether the impacts of low-dose irradiation are harmful or beneficial. These discussions were incited by the Chernobyl catastrophe, after which large areas were polluted with radioactive materials. At that time, it was suggested that perturbed health in irradiated people was caused merely by irradiation phobia.

To move beyond this argument, we initiated investigations using plants: a pea pure line[1] and a natural population of plantain, which is a favorite subject of radioecologists due to its prevalence. The ability to conduct radiation experiments with plant seeds was also important, because seedling apical meristems are a good model of bone marrow stem cells. We used a pea pure line in laboratory experiments and we investigated plantain populations near the Balakovo nuclear power plant.

The State Committee on Protection of the Environment in the Saratov Region and the Ecology Department of the State Saratov University assisted us in choosing plantain populations in a 30-km zone around the nuclear plant and provided information on radiation and environmental conditions in this region. It is important that the nuclear plant performed normally at this time (1998–1999); its fallout in this region, if any, did not enhance natural ground radiation (0.10–0.15 μSv/h).

Cytogenetic,[2] dosimetric and chemical analyses[3] were performed at the Joint Institute for Nuclear Research (JINR). Laboratory studies and investigations of natural populations were analyzed in parallel. In both cases, increased numbers of chromosomal aberrations were registered, which can be understood because low-dose radiation induces chromosomal aberrations that can accumulate and then

---

[1] Seeds were kindly provided by Prof. G. Debelyj, Moscow Institute of Agriculture, Nemchinovka.

[2] Methods were provided by radiobiologists of the laboratory of Prof. V.A. Shevchenko (Vavilov Institute of General Genetics, RAS, Moscow) and the Russian Institute of Agricultural Radiology and Agroecology, RAAS (Obninsk).

[3] Correlations between the chemical pollutants, biological values and nuclear power plant fallout were analyzed.

lead to cell adaptation or death.[4] These results were verified by investigations of bystander effects[5] and hypermutagenesis.[6] We used statistical modelling to study regularities in the appearances of abnormal cells and chromosomes.

Fitting distributions of numbers of abnormalities in seeds and cells revealed combinations of Poisson and geometric statistics.[7] Poisson statistics describe the independent appearances of primary and late damages, and geometric statistics describe selection. Both together describe the processes of adaptation induced by primary impacts, which appears as time-dependent, non-random mutagenesis coupled with selection. These regularities were discussed with Profs. V.B. Priezzhev, G.A. Ososkov (JINR) and other mathematicians. The values of the Poisson and geometric distributions determine their biological interpretation. Thus, we studied the processes of adaptation induced in seed populations and in seedling apical meristems in terms of irradiation dose rates.

Our investigations inspired us, and we began to study problems of statistical modelling of the appearance of abnormal lymphocytes in the blood of individuals living in sites impacted by radiation. We verified that the distribution of individuals on the frequency of abnormal cells can be used to analyze instability processes.[8] We also wished to study instability processes across generations. For our modelling, we chose investigations of blood lymphocytes of individuals living in the settlements of Samburg (Tyumen region) and Maloe Goloustnoe (Irkutsk region), which experienced fallout from nuclear tests (1950) in Novaya Zemlya and Semipalatinsk, respectively. Cytogenetic studies of the blood of individuals living in these sites were collected over several decades at the Novosibirsk Institute of Cytology and Genetics of the Siberian Branch of the Russian Academy of Sciences.

Unfortunately, there are no data concerning cumulative doses by individuals from fallout in Samburg, although a dramatic increase in malignant changes of lungs and digestive organs was documented (investigations provided since 1960). In Maloe Goloustnoe, the accumulated dose by individual from fallout could be 10–40 cSv (an estimate by some authors). At present, ^{137}Cs contaminations are 153 and 118 Bq/kg (lichen and venison, respectively) in Samburg,[9] and 55 Bq/m^2 (soil) in Maloe Goloustnoe.[10]

---

[4]These ideas were based on investigations by Prof. V.I. Korogodin on chromosomal instability (1970) and transcription-regulated mutagenesis (1980).

[5]By Prof. C. Mothersill and Prof. C.B. Seymour.

[6]By Prof. S. Rosenberg.

[7]The methodology and methods of fitting were discussed with Prof. G.A. Ososkov (JINR).

[8]The proof was published in *J. Math. Biol. and Bioinformatics* (in Russian).

[9]Data of the Analytical Centre of the Siberian Branch of the Russian Academy of Sciences, which investigates radionuclide contents in lichen, moss, venison etc.

[10]Dosimetric analyses were performed independently by the radiation monitoring services of Irkutsk, Angarsk, and Novosibirsk, which study radiation effects of nuclear tests in the Semipalatinsk polygon on the South Baikal side.

The modelling clarified some conclusions: time-dependent adaptation proceeds constantly in the environment; it is coupled with instabilities and selection; and their risks can be estimated by statistical modelling. The statistical approach is qualified to investigate the processes induced by the low factors and accompanied by Darwinian selection in different systems. These notions are especially useful to specialists in radiation pollution, ecology, epidemiology, and radiology for the studies of radiation-induced processes. The offered method could be successfully developed by investigators of low-dose effects in other fields.

The ideas and investigations of Prof. Vladimir I. Korogodin were the basis of our adaptation modelling. In addition, we are grateful to several scientists who contributed significantly to these investigations: Prof. V.B. Priezzhev (notions on self-organization of the matter); Prof. G.A. Ososkov (methods of fitting); Prof. J.W. Drake (adaptive mutagenesis and evolution); and Prof. Yu.A. Kutlakhmedov (radioecology notions). We thank Prof. E.B. Burlakova, who supported us at the beginning of our investigations; this was very important. Professors C.Mothersill, G.G. Polikarpov, C.B. Seymour, and A. Yablokov supported our efforts to preserve Nature.

Dubna, Russia   Victoria L. Korogodina

Contents

1	**Introduction**		1
	References		3
2	**Disturbance of Hereditary Material Reserves Is the Main Instrument of Stress**		5
	2.1	Preamble	5
	2.2	Low-Dose-Radiation Effects	6
		2.2.1 Some Regularities of Adaptation in Pre-molecular Investigations	6
		2.2.2 Bystander Effect	8
		2.2.3 Genomic Instability	10
		2.2.4 Transgenerational Response	11
	2.3	Modelling of the Radiation Effects	12
		2.3.1 Probability Models	12
		2.3.2 Theoretical Models of Bystander Effects	13
	2.4	Probability Approach to Risk Assessment of the Chromosomal Instability	15
	2.5	Summary	15
	References		17
3	**Excursus on Statistical Modelling for Population Biology. Statistical Solution of Some Radiobiological Tasks**		23
	3.1	Preliminaries	23
	3.2	Geometric Model of Adaptation	24
	3.3	View on the Structure of Biological Communities	26
		3.3.1 Model of the Geometric Series of Motomura	26
		3.3.2 Fisher's Log-Series	27
		3.3.3 Preston's Lognormal Distribution of Species Abundances	27
		3.3.4 Studies of the Biological Communities That Experienced the Totskij Nuclear Explosion	28
		3.3.5 Influence of the Mutagenic Factor on Cellular Population	28

	3.4	Statistical View on the Cytogenetic Investigations of Instabilities ...	30
		3.4.1 Hypotheses and Models of Appearance of Cells with Abnormalities in Seedlings Meristem of Stress-Irradiated Seeds	30
		3.4.2 Correlative Model of Multiplication of the DNA Damages ..	34
		3.4.3 Hypothesis and Model of the Proliferated Cells Occurrence .	39
		3.4.4 Some Features of Forming and Analysis of Individuals' Distributions on the Number and Frequency of Cells with Abnormalities in Blood Lymphocytes...	41
	3.5	Some Conclusions	51
	3.6	Summary	53
	References		55
4	**Non-linearity Induced by Low-Dose Rates Irradiation. Lab Experiments on Pea Seeds**		57
	4.1	The Seeds of a Plant Are an Important and Convenient Object in Radiobiology and Radioecology	57
		4.1.1 The First Investigations of Non-linearity Were Performed on Plants......................................	58
		4.1.2 Modification of Radioresistance of Plants......................	59
		4.1.3 Meristems Are Critical Tissues of Plants	60
	4.2	Laboratory Experiments on Pea Seeds	62
		4.2.1 On the Object and Methods Used in Lab Investigations	62
		4.2.2 Occurrence of Cells with Abnormalities at Different Dose-Rate Irradiation. Effects of Aging and High Temperature	63
		4.2.3 Non-linearity Induced by Low-Dose Rates Irradiation. Effects of Aging and High Temperature	67
		4.2.4 Adaptive Response of Meristem Cells Induced by Low-Dose Rate Irradiation and Its Combination with High Temperature	69
	4.3	Modelling of Appearance of Cells with Abnormalities	70
		4.3.1 Appearance of Cells with Abnormalities in Seedlings' Meristem	70
		4.3.2 Chromosomal Instability	74
		4.3.3 Multiple Appearances of Abnormal Chromosomes in Meristem Cells. Modelling of Chromosomal Instability	74
		4.3.4 Modification Effects of High Temperature and Aging	77
		4.3.5 Scheme of the Adaptation Process	78
	4.4	Summary	80
	References		81

Contents

5 Adaptation and Genetic Instability in Ecology. Study of the Influence of Nuclear Station Fallout on Plant Populations 83

- 5.1 A View on Natural Communities with Experience in Radiation Impact 84
  - 5.1.1 Consequences of Radiation Accidents 84
  - 5.1.2 Radiation Sources Providing Low-Radiation Stress Comparable with Background 85
- 5.2 Characteristics of Objects and Methods in the Ecological Studies ... 85
- 5.3 Analysis of Averaged Biological Values 89
- 5.4 Scheme of Adaptation Processes in Meristem and Cells 93
- 5.5 Statistical Modelling of the Appearance of Cells with Abnormalities, Abnormal Chromosomes in Cells and Proliferated Cells (PCs) 95
  - 5.5.1 Distributions of Seeds on the Number of Cells with Abnormalities and Cells on the Number of Chromosomes with Abnormalities 95
  - 5.5.2 Distribution of Seeds on the Number of PCs in Seedling Meristem 96
  - 5.5.3 Regularities of the Adaptation-Processes Induced by Nuclear Station Fallout in Plantain Populations 97
- 5.6 Two Evolution Strategies of Survival 101
- 5.7 Risk of Chromosomal Instability 102
- 5.8 Summary 102
- References 104

6 Instability Process Across Generations. Consequences of Nuclear Test Fallout for Inhabitants 107

- 6.1 General Genetic Consequences of Radiation Impacts 108
  - 6.1.1 Consequences of Dramatic Nuclear Impacts 108
  - 6.1.2 Population of the Tundra Nenets Living in the Purov Region of the Yamalo-Nenets Nenets Autonomous Area as a Model to Investigate Consequences of Nuclear Test Fallout Impacts 110
  - 6.1.3 Radiation Effects of the Semipalatinsk Nuclear Test Fallout in the South Baikal Zone (Pribaikal'e) 113
- 6.2 Objects and Methods of Radiation Epidemiology Investigations 115
  - 6.2.1 Description of the Tested Sites and Populations in the Yamalo Nenets Autonomous Area 115
  - 6.2.2 Description of the Tested Sites and Populations in the Pribaikal'e Region 120
  - 6.2.3 Cytogenetic Analysis 121
- 6.3 Statistical Modelling for Persons Living in the Radiation – Polluted Areas 122
  - 6.3.1 Methods of the Statistical Modelling 122

		6.3.2	Examination of the Distributions of the Control Group of Individuals...	122
		6.3.3	Comparison of the Distributions of Individuals with Normal and Low Activated Cells on the Frequency of Cells with Abnormalities in Blood Lymphocytes..	124
		6.3.4	Comparison of the Distribution Structures of Individuals Under 18 and Those Older Living in the Pribaikal'e Region on the Frequency of Cells with Abnormalities	126
		6.3.5	Analysis of Distributions of the Individuals Living in the Siberian Utmost Northern and Pribaikal'e Regions....	128
		6.3.6	Risk of Chromosomal Instability for Individuals	131
		6.3.7	Mechanisms of Genetic Adaptation for Humans..............	132
	6.4	The Appearance of Multi-aberrant Cells Induced by the Radiochemical Industry		134
	6.5	Summary		135
	References...			136
7	**Conclusion**			141
	7.1	Adaptation, Genetic Instability, and Selection Processes		141
	7.2	Regularities of the Adaptation Processes		144
	7.3	Instabilities Induced by Low-Radiation Fallout		145
	7.4	How the Adaptation Process Can Be Presented Statistically		147
	References...			148
8	**Applications**			151
	8.1	Methods of Approximation		151
	8.2	Approximations of Pea Seeds Distributions on the Number of Cells with Chromosomal Abnormalities Irradiated in the Laboratory		154
		8.2.1	Experimental Data on the Pea Seeds Used for Approximations...	154
		8.2.2	Approximation Efficiency of Seed Distributions on the Number of Abnormal Cells	154
		8.2.3	Approximations of Distributions of Pea Cells on the Number of Chromosomal Abnormalities	156
	8.3	Sample of the Calculation of the Synergic Coefficient................		156
	8.4	Approximation of Distributions for Plantain Seeds Growing near the Nuclear Power Plant		158
		8.4.1	Approximations of the Appearance of Cells with Abnormalities in Meristem of Plantain Seedlings	158
		8.4.2	Approximations of the Appearance of Proliferated Cells in Meristem of Plantain Seedlings......................	159

8.5	Analysis of the Correlations of Seed Mortality $(1-S)$ and Parameters of the Plantain Seed Distributions on the Numbers of Proliferated and Abnormal Cells with the Contamination of Chemical Pollutions in Soil (1998, 1999)		162
8.6	Statistical Modelling of the Occurrence Frequency of Cells with Abnormalities in Blood Lymphocytes of Individuals		166
	8.6.1	Experimental Data on Number of Cells with Abnormalities in Blood Lymphocytes of Samples of Individuals Living in Different Regions of Siberia (Table 8.13)	166
	8.6.2	Examination of the Control Samples of Individuals	166
	8.6.3	Examination of the Experimental Data of Individuals Living in the Maloe Goloustnoe and Listvyanka Settlements (Indistinguishability of Distributions for Men and Women)	169
	8.6.4	Approximations of the Experimental Distributions for Samples of Individuals Living in the Settlements of Maloe Goloustnoe and Listvyanka	169
References			171

Glossary 173

Name Index 175

Subject Index 181

Chapter 1
Introduction

Abstract In recent times, many researchers have studied general mechanisms of low-dose radiation effects and have published descriptions of a wide variety of radiation-polluted territories. However, there have not yet been enough analytic investigations of the processes of adaptation. The purpose of this book is to describe those processes of adaptation that are induced by the radiation factor in cells, tissues, and populations. They were examined through a method of statistical modelling that is suitable for analyzing those cases when the radiation factor is comparable with the background. Statistical modelling identifies the processes of the changes' appearance with a statistical description of the fitted system structure on the number of changes. Statistical modelling is an instrument which separates reactions of resistant and sensitive subpopulations. This approach gives us an opportunity to calculate the risks of instability leading to both the accumulation of abnormalities and selection processes. Statistical modelling is a technique that can be successfully developed by investigators of low-dose effects in different fields.

Keywords Aims of investigation • Radiation-induced processes • Adaptation • Statistical modelling • Instability • Darwinian selection

At present, there are vast areas that are polluted by radiation in locations around the world, and the number of such territories is increasing. These territories are especially dangerous because radiation effects continue through generations of living organisms. This reality motivated our research into the adaptation processes. Systematic investigations of the influence of radioactivity on biota were first provided in the 1950s by N.W. Timofeeff-Ressovsky along with his colleagues in his laboratory in the South Urals. At the same time, the famous Timofeeff-Ressovsky school appeared and it significantly contributed to the studies of radiation genetics, radiobiology, and radioecology. We will present several investigations of that kind, although they were published only in the Soviet Union at the time owing to the reality of the cold war.

In the previous two decades, the efforts of radiobiologists were concentrated on investigations of low-dose irradiation, mainly motivated by the Chernobyl accident. Up to the present moment the general regularities of radiation effects have been studied: it has been shown that a low radiation factor induces the bystander effect (Mothersill and Seymour 2001), genomic (Morgan 2003a, b) and transgenerational (Dubrova 2003) instability. Microbeam technology allows one to demonstrate that targeted cytoplasm irradiation results in mutations in the nuclei of hit cells and the presence of non-targeted effects (Prise et al. 1998; Prise 2006). Recent studies have shown that irradiation alters epigenetic characteristics which regulate gene expression in directly exposed tissues as well as in distant bystander tissues (Kovalchuk and Baulch 2008). Radiation ecologists and epidemiologists have collected data on low-dose radiation effects by monitoring polluted areas (Yablokov et al. 2009). Nowadays it is clear that the effects of radiation impact can continue in populations for some decades (Yablokov et al. 2009; Shevchenko et al. 1992; Tanaka et al. 2006).

The above investigations have discovered the mechanisms of the bystander effect and genomic instability induced by low-dose radiation, while the others relate to the monitoring of polluted areas. The purpose of our investigation is to describe and analyze these radiation effects as the processes induced in cells, tissues, and populations. It is the adaptation processes that always take place in living matter. This viewpoint helps us to understand the nature and factors of the induced processes, as well as to determine the characteristics of observed radiation effects and their limitations.

Such investigations can be performed by means of statistical modelling on the basis of distribution of the number of abnormalities. Statistical modelling is a delicate instrument which separates the reactions of resistant and sensitive subpopulations. It allows one to consider the processes induced by low-dose radiation when radiation intensity does not exceed the background. This approach gives us an opportunity to calculate the risks of instability leading to both the accumulation of abnormalities and selection processes.

The developed method is qualified to investigate the processes induced by the low factors and accompanied by Darwinian selection in different systems. Its idea consists of the following: stress factors induce a cascade of successive random changes in biological systems. Such processes are always accompanied by Darwinian selection, which eliminates variants unfit for the new conditions. The statistical modelling identifies the processes of the changes' appearance with the statistical description of the fitted system structure on the number of changes. It appears that this statistical description is specific, and parameters of distributions can be used to study the characteristics of fit processes and system resistance. This book includes experimental studies, theoretical investigations and examples of calculations that enable one to use statistical modelling in practice.

This book has eight chapters, including an Introduction (Chap. 1). The second chapter (Chap. 2) presents a brief review of the phenomenology and mechanisms of low-dose radiation effects. The consequences of the background-level radiation tend to provoke many debates, and the mathematical bases of the adaptation model (Florko and Korogodina 2007; Florko et al. 2009) are shown (Chap. 3) to verify the conclusions. The lab studies of pea seeds are presented (Chap. 4) to demonstrate the general radiation effects induced by the precisely measured low irradiation (Korogodina et al. 2005). In Chap. 5, the presented ecological investigations show the process of adaptation to radiation fallout in natural plantain populations growing in the 30-km zone of an operating nuclear power station (Korogodina et al. 2010). Chapter 6 presents studies on the blood lymphocyte cells of individuals living in territories which experienced nuclear test fallout more than 50 years ago and which have demonstrated radiation-induced chromosomal instability across four generations (Korogodina et al. 2010). The Conclusion (Chap. 7) formulates some general principles of the adaptation model and its consequences. The Application (Chap. 8) contains a description of methods, approximations, and calculations.

First of all, this book is addressed to specialists in radiation pollution, ecology, epidemiology, radiology, and also social sciences. The method offered can be successfully developed by investigators of low-dose effects in other fields. It contains approaches to solving the problem of correct fitting for small statistics, which is an essential part of such research. The book presents pioneering and valuable investigations in this field in Russia which were long unknown to Western scientists.

References

Dubrova YE (2003) Radiation-induced transgenerational instability. Oncogene 22:7087–7093

Florko BV, Korogodina VL (2007) Analysis of the distribution structure as exemplified by one cytogenetic problem. PEPAN Lett 4:331–338

Florko BV, Osipova LP, Korogodina VL (2009) On some features of forming and analysis of distributions of individuals on the number and frequency of aberrant cells among blood lymphocytes. Math Biol Bioinform 4:52–65 (Russian)

Korogodina VL, Florko BV, Korogodin VI (2005) Variability of seed plant populations under oxidizing radiation and heat stresses in laboratory experiments. IEEE Trans Nucl Sci 52: 1076–1083

Korogodina VL, Florko BV, Osipova LP (2010) Adaptation and radiation-induced chromosomal instability studied by statistical modeling. Open Evol J 4:12–22

Kovalchuk O, Baulch JE (2008) Epigenetic changes and nontargeted radiation effects – is there a link? Environ Mol Mutagen 49:16–25

Morgan WF (2003a) Non-targeted and delayed effects of exposure to ionizing radiation: I. Radiation-induced genomic instability and bystander effects *in vitro*. Radiat Res 159: 567–580

Morgan WF (2003b) Non-targeted and delayed effects of exposure to ionizing radiation. II. Radiation-induced genomic instability and bystander effects *in vivo*, clastogenic factors and transgenerational effects. Radiat Res 159:581–596

Mothersill CE, Seymour CB (2001) Radiation-induced bystander effects: past history and future perspectives. Radiat Res 155:759–767

Prise KM (2006) New advances in radiation biology. Occup Med (Lond) 56(3):156–161

Prise KM, Belyakov OV, Folkard M et al (1998) Studies of bystander effects in human fibroblasts using a charged particle microbeam. Int J Radiat Biol 74:793–798

Shevchenko VA, Pechkurenkov VL, Abramov VI (1992) Radiation genetics of the native populations. Genetic consequences of the Kyshtym accident. Nauka, Moscow

Tanaka K, Iida S, Takeichi N et al (2006) Unstable-type chromosome aberrations in lymphocytes from individuals living near Semipalatinsk nuclear test site. J Radiat Res (Tokyo) 47(Suppl A):A159–A164

Yablokov AV, Nesterenko VB, Nesterenko AV (2009) Chernobyl: consequences of the catastrophe for people and the environment. Ann N Y Acad Sci 1181:vii–xiii, 1–327

Chapter 2
Disturbance of Hereditary Material Reserves Is the Main Instrument of Stress

Abstract This review presents data on the low-dose radiation effects and their mathematical models. Its purpose is to demonstrate that radiation stress leads to processes of instability that can be revealed as different phenomena. The phenomena of radioadaptation, nonlinear response induced by low-dose irradiation, hormetic effect, and continued instability across generations, and stimulation of proliferation are considered. Our special interest is the investigation of the bystander effect which clarified some of these phenomena. The regularities of the bystander effect, genomic and transgenerational instability are considered. The modelling of these radiation effects is discussed: the models offered by Yu.G. Kapultsevich (probabilistic), D.J. Brenner et al. ("Bystander and Direct"), H. Nikjoo and I.K. Khvostunov (Diffusion model), and B.E. Leonard ("Microdose Model") are presented. The investigations of Russian scientists and the Timofeeff-Ressovsky school are presented as being of special interest to Western scientists owing to this information not having been published in the West due to the cold war.

Keywords Low-dose irradiation • Radioadaptation • Nonlinear response • Hormetic effect • Stimulation of proliferation • Bystander effect • Genomic instability • Transgenerational instability • Probabilistic model • Bystander and Direct model • Diffusion model • Microdose Model

2.1 Preamble

In 1939 N.W. Timofeeff-Ressovsky published the fundamental ideas of the microevolution process (Timofeeff-Ressovsky 1939). He considered dramatic fluctuations of genotype material in populations leading to the high mortality of organisms as essential equipment for evolution because a new genotype can only be multiplied at low concentrations of the existing ones.

Such a situation is realized by a stress factor: its mechanisms rapidly increase the number of new mutations in the population coupled with a strong selection of the fitted genotype. The regularities of the stress response are the same for any live system, although the details can alter at different levels of life organization. This chapter considers the disturbance of hereditary material reserves[1] by stress as its main instrument, and some principles of adaptation will be demonstrated.

2.2 Low-Dose-Radiation Effects

2.2.1 Some Regularities of Adaptation in Pre-molecular Investigations

History of the investigations of low-dose-radiation effects: In the mid-1950s, investigations of low-radiation biological effects were organized in the USSR at the South Urals laboratory by Nikolay W. Timofeeff-Ressovsky (Timofeeff-Ressovsky and Tyuryukanov 2006; Timofeeff-Ressovsky and Timofeeff-Ressovskaya 2006). N.W. Timofeeff-Ressovsky and his laboratory researchers N.V. Luchnik, N.A. Poryadkova, E.I. Preobrazhenskaya, and others described a nonlinear dependence of cytogenetic effects on the radiation dose (Timofeeff-Ressovsky et al. 1950–1954; Timofeeff-Ressovsky 1956; Luchnik 1958) (the proportional dependence of mutation frequency on dose irradiation had been established earlier (Timofeeff-Ressovsky et al. 1935)). This was the first biological description of the non-linear effects induced by the low-radiation factor. From the 1960s on, the research by N.W. Timofeeff-Ressovsky and his colleagues was continued at the Russian Institute for Medical Radiology in Obninsk where the famous Timofeeff-Resovsky radiobiological school was established (Korogodin 1993).

The phenomenon of radioadaptation: In the mid-twentieth century, anthropogenic radioactive contamination of territories became a significant factor in ecology, and adaptation of natural populations to irradiation was systematically studied. In the USSR these investigations were mainly performed at the Russian Institute of Agricultural Radiology and Agroecology (Obninsk) and the N.I. Vavilov Institute of General Genetics (Moscow). The scientists of these institutes studied the influence of an artificially high radiation background on the natural plant populations (Cherezhanova and Alexakhin 1971) in the zone of the Kyshtym and Chernobyl accidents (Shevchenko et al. 1999).

The analysis of the consequences of the Kyshtym radiation accident in the South Urals (1957) showed a high variability of cytomorphological and physiological characteristics in plants (reviewed in (Pozolotina 1996)). The scientists described the plant radioresistance increasing under chronic irradiation. This phenomenon was called "radioadaptation". V.A. Shevchenko suggested that the population divergence

---

[1] Here we consider the changes of cells and chromosomes.

2.2 Low-Dose-Radiation Effects

on radiation resistance was related to changes in the repair system efficiency (Shevchenko et al. 1999).

These results agree with the increased mutation frequency in barn swallows (Ellegren et al. 1997) and in wheat upon chronic exposure to the ionizing radiation produced by the Chernobyl accident (Kovalchuk et al. 2000, 2003).

The nonlinear response induced by low-dose irradiation: In the mid-1990s, E.B. Burlakova published the results of the experiments in Russian journals, which showed the following: the low-dose-radiation effect has a non-monotonic character; there is an inverse relation of the low-dose effect on radiation intensity; the response depends on the initial characteristics of biological objects; low-dose-rate radiation is more effective than acute in some intervals (Burlakova 1994; Burlakova et al. 1996, 1999, 2000). The hypothesis suggested that the repair systems induced by low-dose radiation differed from those that worked at sub-lethal doses (Burlakova 1994). Natural investigations have shown increased radiation efficiency in populations polluted due to the Chernobyl (Geras'kin et al. 1998) and Kyshtym (Shevchenko et al. 1992) accidents.

Both hypersensitivity and radioresistance was also demonstrated on mammalian cells that were explained by the hypothesis of inducible repair systems (overviewed in (Averbeck 2010)). It was shown that stress factors activate repair systems, the operation of which results in an accumulation of abnormalities (Longerich et al. 1995).

At present, the nonlinear effects are considered to be the result of the bystander effects and genomic instability coupled with the stress-induced specific repair systems (Mothersill and Seymour 2005; Averbeck 2010). Although the nonlinear dose-effect is described for different objects, it remains discussible (Zyuzikov et al. 2011; Little 2010). One of the debatable questions is a linear no-threshold concept linked with the problem of radiation risks assessment (Vaiserman et al. 2010).

Hormetic effect: In the 1970s, data were published on the stimulation effect of low-dose radiation. The so-called radiation-induced hormetic effects were discovered in many branches of life sciences (Luckey 1980; Kuzin 1993; Petin et al. 2003) including protection against spontaneous genomic damage (Feinendegen 2005). This phenomenon can be explained by different reasons.

Hormesis induced by low-dose ionizing radiation often reflects stimulation of cell proliferation (Gudkov 1985; Liang et al. 2011).

It was found that low-dose radiation induces an "adaptive response", which implies a pre-treatment of cells with a low-radiation dose followed several hours later by exposure to a much higher dose (Upton 2001). The adaptive response can reduce radiation-induced DNA damage, mutagenesis, and the frequency of chromosomal aberrations, micronuclei and cell transformants. In (Rigaud and Moustacchi 1996), the authors reviewed the experimental results showing that a prior exposure to a low dose of ionizing radiation induces an adaptive response expressed as a reduction of mutation in various cell systems. The published review (Mothersill and Seymour 2004) considers how bystander effects may be related to observed adaptive responses. In the authors' opinion, it is possible that low-dose exposures cause removal of cells carrying potentially problematic lesions, prior to radiation exposure. Then, the adverse, adaptive or apparently beneficial response will be related to the background damage carried by the original cell population.

Continued instability across generations: As early as 1925, G.A. Nadson and G.S. Filippov (1925) described the mutagenic action of radium on mold, fungi and yeasts. Shortly thereafter, these authors reported the *"en masse"* appearance of morphologically-diverse colonies and cells in the progeny of irradiated yeasts (Nadson and Filippov 1932). These studies were stopped in the USSR after Filippov's death from tuberculosis in 1934 and Nadson's arrest (1937) and execution in 1939. In France, studies of this kind were started by Lacassagne and co-authors (1939), but were terminated due to World War II.

At the end of the 1960s, V.I. Korogodin and his colleagues studied the so-called "cascade mutagenesis" that is a continuous (several hundreds of cell generations) appearance of new races (phenotypic variants) in individual unstable clones formed with high frequency in diploid yeasts after ionizing radiation or any other mutagenic treatment (Korogodin and Bliznik 1972). It was shown that cascade mutagenesis was induced by primary sub-lethal lesions. The phenomenon is that the mutation process, caused by a single irradiation act, occurs at a greater rate, much more than 10^{-2} mutations per cell per division. In diploid cells, the effects of several primary lesions may be summarized and inherited resulting in various mitotic disorders (Korogodin et al. 1977). It was also shown that 60 % of new races were originated in the shoulder region of the survival curve (Korogodin et al. 1972), and their number increased under non-optimal conditions of cultivation (Bliznik et al. 1974). From 1972 to 1977, Korogodin and his colleagues showed the relation of this phenomenon to chromosomal instability (a series of articles published in the Russian journal "Radiobiologiya", Vs. 12–17).

Proliferation: It was shown on human lymphocytes that cells at different stages of the cell cycle differ in sensitivity (Boei et al. 1996). Thus, the cellular population consists of proliferated radiosensitive and radioresistant subpopulations, and resistant resting cells. The heterogeneity of cells results in their different proliferative activity (Gudkov 1985) and plays an important role in the preservation of a proliferated cell pool.

The low-dose radiation activates division within the resting cells. For the first time, Nikolay V. Luchnik published data on the stimulation of resting cells leading to proliferation in plants (1958). He noticed that this mechanism was not compensatory with respect to the cells' killing, but independent.

2.2.2 Bystander Effect

History: A bystander effect was described by W.B. Parsons et al. (1954), who studied factors which induced chromosomal damages (clustogenic factors) registering in the blood of irradiated patients as far back as 1954. By the description of the

2.2 Low-Dose-Radiation Effects

authors, the bystander effect was observed in the cells, which were non-irradiated but still behaved the same as the irradiated ones: they died or demonstrated genomic instability. In 1968, the data of such effects were described by other scientists (Goh and Sumner 1968). In 1997, the hypothesis was offered by C.B. Seymour and C. Mothersill that the signal (or factor) produced in a medium by an irradiated cell was able to induce genomic instability-type effects in a distant progeny (Seymour and Mothersill 1997). At present, the bystander effect induction is well-documented at the low dose of soft and dense radiation (Little et al. 1997; Schettino et al. 2003), both *in vitro* (Morgan 2003a) and *in vivo* (Morgan 2003b).

The standard paradigm of the radiation biological effect ("target theory") lies in the fact that biologically-meaningful consequences of that effect are connected with DNA damage (Timofeeff-Ressovsky and Zimmer 1947). Particularly, double strand breaks of DNA are the reason for mutations, transformation of cells and their death (Pfeiffer 1998). In the 1990s, many reports on "non-targeted" effects were published. But these effects were not the result of the direct radiation effect for DNA. The authors observed abnormalities (chromatid exchanges (Nagasawa and Little 1992; Little 2000), chromosomal aberrations (Little 2000; Lorimore and Wright 2003), apoptosis (Mothersill and Seymour 1997, 2000), formation of the micronuclei (Prise et al. 1998; Sedelnikova et al. 2007), cells' transformation (Sigg et al. 1997), mutations (Zhou et al. 2000) and the gene expression changes (Hickman et al. 1994; Mothersill and Seymour 2001)) in non-irradiated cells, which were neighbors of the irradiated ones.

The bystander effect was manifested through the use of several different methods: studies of the effect were performed *in vitro* by a microbeam at the Cancer Institute (UK), when the fixed separate cells were damaged (Sawant et al. 2001; Belyakov et al. 2002); in another instance, while transferring the non-irradiated cells into the media where irradiated cells were cultivated, the effects were registered among the non-irradiated cells (Mothersill and Seymour 1997, 1998; Nagar et al. 2003); and it was also shown to occur by means of the probability approach which is based on irradiation of the cells' culture at the minute particle flux (Nagasawa and Little 1992; Deshpande et al. 1996). The main investigations were devoted to the death of cells, mutations and chromosomal aberrations (Mothersill and Seymour 2001). The bystander effect phenomenology was developed and published in (Mothersill and Seymour 2001; Morgan 2003a, b; Little and Morgan 2003; Lorimore and Wright 2003; Lorimore et al. 2003).

Dependence on the dose and dose-rate radiation: The authors (Schettino et al. 2003) studied the bystander effect induced by ultra-soft X-rays in V79 cells. The linear-square law of the cell survival dependence on dose irradiation and hypersensitivity were observed at low-dose radiation. The analysis of the distance between the non-irradiated damaged cells and non-damaged cells showed the clusters of bystander-damaged cells.

The bystander effect depends non-linearly on the dose irradiation (Hickman et al. 1994; Deshpande et al. 1996; Prise 2006), and its manifestation is maximal at low doses (Belyakov et al. 2000). Some authors assume that induction of the bystander effect has a threshold (Schettino et al. 2005; Liu et al. 2006).

Signaling: induction, transference and reception: It is shown that irradiated mammalian cells can generate and transmit signals to the non-irradiated neighbors involving reactive oxygen species and nitric oxide species (Shao et al. 2006, 2008; Kashino et al. 2007; Portess et al. 2007). There are different data concerning the repair-deficient influence on the bystander responses (Mothersill et al. 2004; Nagasawa et al. 2003).

C. Mothersill and C.B. Seymour assume that small proteins and peptides can be signaling carriers from the irradiated cell to the non-irradiated one (2001); in some cases, direct cell contact is needed, and in others, it is not (Mothersill and Seymour 2002). It was shown that reactive oxygen species, such as superoxide and hydrogen peroxide, and calcium signaling are important modulators of bystander responses (Lyng et al. 2006).

2.2.3 Genomic Instability

Studies of the bystander effects showed their involvement in the demonstration of genomic instability. Genomic instability is characterized by increased changes in the genome, and manifests itself as chromosomal aberrations, genetic mutations and amplifications, late cell death, aneuploidy induction of micronuclei, and microsatellite instability (Watson et al. 2000; Little 2000; Mothersill and Seymour 2001; Little and Morgan 2003; Lorimore and Wright 2003; Lorimore et al. 2003; Kovalchuk and Baulch 2008; Averbeck 2010). Radiation induces genomic instability in the irradiated cell at delayed times after irradiation and in the progeny of the irradiated cell (Little 2000). Such instability may be a prerequisite for cancer. Genomic instability was observed *in vitro* (Morgan 2003a, 2011) and *in vivo* (Watson et al. 2000; Morgan 2003b, 2011).

Genomic instability was documented in blood samples from accidentally irradiated individuals, individuals after radiotherapy, survivors of Hiroshima and Nagasaki in 1977, liquidators and children of Chernobyl, human blood samples irradiated *in vitro*, and patients with cancer predisposition syndromes (Bloom's syndrome, Fanconi's anemia, Xeroderma pigmentosum) (Emerit et al. 1997).

The role of the dose and radiation quality: Genomic instability was observed both at high and low linear energy transfer (LET) (Aypar et al. 2011). But investigations have shown the LET-dependent induction and expression of genomic instabilities (Okada et al. 2007; Aypar et al. 2011).

To investigate the long-term biological effect of extreme low-dose ionizing radiation, the authors (Okada et al. 2007) irradiated normal human fibroblasts with carbon ions and gamma-rays. These studies have indicated that high-LET radiation (carbon ions) causes the effects which differ from those induced by low-LET radiation (gamma-rays), and that a single low dose of heavy ion irradiation can affect the stability of the genome of many generations after exposure. The experimental results published in (Kadhim et al. 2006) assume that the dose might be the most significant factor in determining induction of genomic instability after low-LET radiation. In (Aypar et al. 2011) the authors have tested the hypothesis that irradiation induces epigenetic aberrations, which could eventually lead to genomic instability, and that the epigenetic aberrations induced by low-LET radiation differ from those induced by high-LET irradiations.

The authors (de Toledo et al. 2011) emphasize that the radiation dose and radiation quality (LET) are very important in determining the nature of the induced effect. Radiation type, dose rate, genetic susceptibility, cellular redox environment, stage of cell growth, level of biological organization and environmental parameters are the factors which modulate interactions among signaling processes and determine short- and long-term outcomes of low-dose exposures.

2.2.4 Transgenerational Response

History: Enhanced genetic changes continuing in descendants were observed and investigated by G.A. Nadson, B. McClintock, and C. Auerbach many years ago (Nadson and Filippov 1932; McClintock 1938; Auerbach and Kilbey 1971). These famous scientists considered radiation as an inductor of genomic instability (Nadson and Filippov 1925, 1932; Auerbach and Kilbey 1971; McClintock 1984). C. Auerbach obtained evidence of an increased mutation rate in the first-generation (F1) Drosophyla offspring of exposed its parents (Auerbach and Kilbey 1971). In the 1970s V.I. Korogodin and his colleagues investigated regularities of "cascade mutagenesis" on yeasts, when appearance of new races (phenotypic variants) continued in several hundreds of cell generations (Korogodin et al. 1977). Luning et al. investigated the frequency of dominant lethal mutations in the germline of non-exposed offspring of irradiated male mice, and obtained evidence for transgenerational destabilization of the genome (Luning et al. 1976).

Transgenerational instability: In recent years, evidence has been obtained for the induction of persistent elevated levels of mutation rates in the progeny of irradiated cells both *in vivo* (Morgan 2003a) and *in vitro* (Morgan 2003b). Dubrova et al. (2000) showed elevated minisatellite mutation rates in the mouse germline by low-dose chronic ionizing irradiation, and a potential contribution of genomic instability to transgenerational carcinogenesis has been assumed (Dubrova 2003).

Many authors have reported that the offspring of irradiated parents demonstrate transgenerational instability, and the number of affected offspring may substantially exceed the one predicted by the target theory (Luning et al. 1976; Wiley et al. 1997; Barber et al. 2006). The long-term destabilization of the genome could also lead to cancer. In (Vorobtsova et al. 1993), the incidence of cancer in the offspring of irradiated male mice painted with acetone or with acetone solution was analyzed. The authors showed an elevated incidence of cancer among the offspring of the irradiated males. The studies of genetic instability across generations of low-dose-rate irradiated mice which indicated the genetic instability in the F1, F2, and F3 generations from the irradiated males were published in (Zaichkina et al. 2009).

Transgenerational response is an attribute of the genome-wide destabilization: Investigations performed by Yu.E. Dubrova, R.C. Barber and their colleagues have shown that the expanded simple tandem repeat (ESTR) mutation rate in DNA samples extracted from the germline (sperm) and somatic tissues taken from the F1 offspring of male mice, exposed to 1 or 2 Gy of acute X-rays, are equally elevated in both cases (Barber et al. 2006). Transgenerational changes in somatic mutation rates were observed by studying the frequency of chromosomal aberrations (Vorobtsova 2000), micronuclei (Fomenko et al. 2001) and lacI mutations (Luke et al. 1997) in the F1 offspring of irradiated male mice and rats. The referred data have demonstrated that exposure to ionizing radiation results in the induction of a transgenerational signal in the germline of exposed parents which can then destabilize the genomes of their offspring. Experimental studies on the transgenerational effects of post-Chernobyl paternal irradiation were performed and showed an elevated frequency of chromosomal aberrations among the children of exposed fathers (Aghajanyan et al. 2011).

Y.E. Dubrova (2012) analyzed the spectrum of delayed mutations resulting from the ongoing instability, and demonstrated the difference from those resulting from direct induction, which, at the same time, were very close to the spectrum of spontaneous mutations. This means that radiation-induced genomic instability may result from enhancement of the process of spontaneous mutation, and a genome-wide destabilization of the F1 genome could be attributed to replication stress.

Epigenetic changes and non-targeted radiation effects: The data published indicated that radiation-induced genomic instability, bystander, and transgenerational effects may be epigenetically mediated (Nagar et al. 2003; Kaup et al. 2006; Jirtle and Skinner 2007; Aypar et al. 2011). The epigenetic regulation of gene expression includes DNA methylation, histone modification, and RNA-associated silencing (Jaenisch and Bird 2003). Recent studies have demonstrated that ionizing radiation exposure changes epigenetic parameters in directly exposed tissues and in distant bystander tissues; transgenerational radiation effects were also proposed to be of an epigenetic nature (reviewed in (Kovalchuk and Baulch 2008)).

2.3 Modelling of the Radiation Effects

2.3.1 Probability Models

In 1935, N.W. Timofeeff-Ressovsky, K.G. Zimmer and M. Delbrük (Timofeeff-Ressovsky et al. 1935) published the paper "Über die Nature der Genmutation und der Genstruktur", in which they described quantitatively the dependence of gene mutation frequency on the dose irradiation in Drosophila. The "hit principle" and "target theory" were formulated in monographs by D.E. Lea (1946) and N.W. Timofeeff-Ressovsky and K.G. Zimmer (1947) in 1947–1948. The model postulated a single hit in a single target as a "triggering event" and explained the exponential form of the survival curve for single-hit events, i.e., the survival of irradiated viruses.

The sigmoid form of the curve "dose-effect" was investigated by Y.G. Kapultsevich and his colleagues (Kapultsevich and Petin 1977; Kapultsevich 1978). The Kapultsevich model postulates that the registered effect in cells is the result of a number of discrete damages caused by hitting in intracellular structures ("the targets"). The number of such damages is Poisson-distributed in cells. The authors studied the probability of cell division depending on the number of damaged targets. This probability model describes quantitatively the regularities of cell division in a wide interval of sub-lethal-dose irradiation (Kapultsevich 1978). This model doesn't consider the time-dependence of cell damaging.

2.3.2 Theoretical Models of Bystander Effects

Single-cell irradiators and new experimental assays are rapidly expanding the ability to detect a subtle biological phenomenon such as the bystander effect. Mathematical models are needed to interpret the results of the microbeam and low-dose experiments. Although there is increasing evidence that the bystander effects play an important role in the low-dose-radiation response, some models have been developed to account for these phenomena.

"Bystander and Direct" model: Brenner et al. (2001; Brenner and Sachs 2002) offered the so-called "Bystander and Direct" (BaD) model to explain the results of the α-particle microbeam experiments. The authors postulated a bimodal character of the bystander effect: "all" or "nothing" in the cells of a small bystander-hypersensitive subpopulation. The cells of this subpopulation are also sensitive to a direct hit of particles which inactivates cells. For α-particles the whole effect (Y) summarizes actions of the direct hit (D) and bystander effect (B): $Y = B + D$.

Cells that are directly damaged emit signals to the nearest cells that are not. The model doesn't consider the nature of the signal (k), which is the model parameter. Let us suggest that if the α-particle hits the cell, then any of the nearest hypersensitive k cells will be activated.

Non-hypersensitive cells can be activated only through a direct hit by the α-particle. Therefore, statistics of these events are a standard Poisson law. Its sample mean can be determined by a formula $D = \mu N$, where N is an averaged number of particles per cell nucleus, and μ is the other model parameter. Thereby, the model has two parameters.

These premises make it possible to describe satisfactorily the data presented in (Brenner et al. 2001). It is shown that an increase of exposure time at a constant irradiation dose leads to the inverse dose-rate-effect that is the increasing of the Y effect at the decreasing of the dose-rate-irradiation. The analysis has shown that the bystander effect is observed only at low doses of ~0.2 cGy and less. At the lower doses, the bystander effect can dominate. If we are to extrapolate the risk from the sub-lethal dose interval where the direct hitting effects dominate, the underestimation of low-dose-radiation risk is possible.

Diffusion model: H. Nikjoo and I.K. Khvostunov (2004) developed a model of the radiation-induced bystander effect based on the diffusion principle of signal transferring. The Bystander Diffusion Model (BSDM) assumes that a low-molecular-weight protein can be emitted as the signal carrier from the damaged cell and diffuses in intercellular media to the undamaged cell. Cell inactivation and induced oncogenic transformation by the microbeam and broadbeam irradiation systems were considered. The model postulates that the bystander response observed in non-hit cells originates from specific signals received from inactivated cells. The bystander signals are assumed to be protein-like molecules spreading in the culture media by Brownian motion. The bystander signals are supposed to switch cells into a state of cell death (apoptotic/mitotic/necrotic) or induced oncogenic transformation modes. The model predictions for cell inactivation and induced oncogenic transformation frequencies closely correspond to the observed data from microbeam and broadbeam experiments.

The model can be used to explain the survival of cells studied in experiments on transferring the unirradiated cells into the irradiated cell cultural media. This model is also applicable to interpreting the dose-effect curve for survival and carcinogenic transformation of the cells irradiated by alpha-particles. In the case of irradiation with a constant fraction of cells, the transformation frequency for the bystander effect increases with the increase of the radiation dose. The BSDM predicts that the bystander effect cannot be interpreted solely as a low-dose-effect phenomenon. It is shown that the bystander component of the radiation response can increase with the dose and can be observed at high doses as well as low doses (Nikjoo and Khvostunov 2004). The validity of this conclusion is supported by the analysis of experimental results from the high-LET microbeam experiments.

A composite microdose adaptive response and bystander effect model: It is known that an adaptive response may reduce risks of adverse health effects due to ionizing radiation. But very low-dose bystander effects may impose dominant deleterious human risks. These conflicting effects contradict the linear no-threshold human risk model. The dose and dose-rate-dependent microdose model, which examines adaptive response behavior, was described by B.E. Leonard in (Leonard 2008a). The purpose of this work was to obtain new knowledge regarding adaptive response and bystander effects, and illustrate the use of the model for planning radiobiological experiments.

In (Leonard 2008a, b), the author published his work, which provides a composite, comprehensive Microdose Model (MM) that is also herein modified to include the bystander effect. The MM describes the biophysical composite adaptive response and bystander effects and quantifies the accumulation of hits (Poisson distributed, microdose specific energy depositions) to cell nucleus volumes. This model gives predictions of the dose response at very low-dose bystander effect levels, higher-dose adaptive response levels and even higher-dose direct (linear-quadratic) damage radiation levels. The author found good fits of the model to both bystander effects data from the Columbia University microbeam facility and combined adaptive response and bystander effects data for low- and high-LET data.

The five features of major significance provided by the Microdose Model (Leonard 2008a) so far are: (1) single specific energy hits initiate adaptive response; (2) mammogram and diagnostic X-rays induce a protective bystander effect as well as adaptive response radioprotection; (3) for mammogram X-rays, the adaptive response protection is retained at high primary dose levels; (4) the dose range of the adaptive response protection depends on the value of the specific energy per hit; (5) alpha-particle-induced deleterious bystander damage is modulated by low-LET radiation.

2.4 Probability Approach to Risk Assessment of the Chromosomal Instability

Apparently the data mentioned above can be interpreted in different terms, including the "harmful" or "useful" aspects of the phenomena (Burlakova et al. 1996; Mothersill and Seymour 2005). It is useful to consider the influence of low-dose radiation in view of the nature law – adaptation (Korogodina et al. 2010). It is simple to analyze the adaptation processes using chromosomal instability, which could be an indicator of the bystander effect (Little et al. 1997; Lorimore et al. 2003).

The usual biological methods are based on averaged experimental values and give rough estimations of the chromosomal instability risks (Shevchenko et al. 1992). Contrarily, the risk assessments given by theoretical modelling are based on theoretical assumptions (Brenner et al. 2001; Brenner and Sachs 2002; Nikjoo and Khvostunov 2004; Leonard 2008a). The statistical view is based on the experimental data and allows one to consider resistant and sensitive fractions which follow different regularities. The structure of distributions of organisms on the number of abnormalities reflects the characteristics of the adaptation processes: the dose rate of the primary injuring, and intensities of the bystander processes and repair (Florko and Korogodina 2007; Korogodina et al. 2010). Statistical modelling can be used to investigate fundamental problems of microevolution processes and the risk assessment of genomic instability in ecology, epidemiology, medicine, and cosmic investigations.

2.5 Summary

The study of low-dose radiation effects began in the middle of the twentieth century. The Urals' N.W. Timofeeff-Ressovsky laboratory found the non-linear regularities (Timofeeff-Ressovsky et al. 1950–1954; Timofeeff-Ressovsky 1956; Luchnik

1958). Then, other low-radiation effects were investigated: hypersensitivity and elevation of radioresistance observed in radiation-polluted territories (Shevchenko et al. 1992), the stimulation of cells to divide (Luchnik 1958), the induction of adaptive response (Rigaud and Moustacchi 1996) and radiation hormesis (Luckey 1980) that protect cells against genomic damages, and instability across generations (Korogodin et al. 1977).

A new step in the investigations was established when the bystander effect was revealed.

Non-targeted effects: A bystander effect was described by W.B. Parsons et al. in 1954. In the 1990s, many reports of "non-targeted" effects were published. The authors observed abnormalities (chromatid exchanges, chromosomal aberrations, apoptosis, formation of the micronuclei, cells' transformation, mutations and gene expression changes) in non-irradiated cells, which were neighbors of the irradiated ones (reviewed in (Mothersill and Seymour 2001, 2006)). It was shown that the bystander effect depends non-linearly on the dose irradiation (Prise 2006). In 1997, the hypothesis was offered by C.B. Seymour and C. Mothersill that the signal (or factor) produced in a medium by an irradiated cell was able to induce genomic instability-type effects in a distant progeny (Seymour and Mothersill 1997).

The bystander effect is linked to the phenomenon of the radiation-induced genomic instability that manifests itself as chromosomal aberrations, genetic mutations, late cell death, and aneuploidy (Kovalchuk and Baulch 2008). The genomic instability was observed *in vivo* and *in vitro* (Morgan 2003a, b). Both non-targeted phenomena include intra- and intercellular signaling, involving reactive oxygen species (Averbeck 2010).

Transgenerational response: In recent years, evidence has been obtained for the induction of persistent elevated levels of mutation rates in the progeny of irradiated cells (Morgan 2003b). Dubrova et al. (2000) showed the elevated minisatellite mutation rates in the mouse germline induced by low-dose chronic ionizing irradiation. Experimental studies on the transgenerational effects of post-Chernobyl paternal irradiation were performed and showed the elevated frequency of chromosomal aberrations among the children of exposed fathers (Aghajanyan et al. 2011). These data have demonstrated that exposure of the individuals to ionizing radiation results in the induction of a transgenerational signal in the germline of exposed parents which can then destabilize the genomes of their offspring. The inheritable radiation-induced genomic instability in all F1 offspring is assumed to be an epigenetic type of transmission (Kovalchuk and Baulch 2008).

Modelling of the radiation effects: Some models were considered here. The first one was the probability model by Yu. G. Kapultsevich (1978). In this model, the experimental data were used. Such modelling helps to understand general regularities.

The theoretical models describe the processes based on the premises of scientists, and then the mathematical conclusions are compared with the experimental data. Such theoretical models were developed by D.J. Brenner ("BaD") (Brenner et al.

2001), B.E. Leonard (Microdose model) (Leonard 2008a, b), and H. Nikjoo and I.K. Khvostunov (Diffusion model) (2004). The mathematical modelling is aimed at studying some concrete problems.

Risk assessment: Statistical modelling is based on experimental data and allows one to study resistant and sensitive fractions (Florko and Korogodina 2007; Korogodina et al. 2010). It can be used to investigate the characteristics of microevolution processes and the risks of genomic instability in ecology, epidemiology, medicine, and cosmic investigations.

References

Aghajanyan A, Kuzmina N, Sipyagyna A et al (2011) Analysis of genomic instability in the offspring of fathers exposed to low doses of ionizing radiation. Environ Mol Mutagen 52:538–546

Auerbach C, Kilbey BJ (1971) Mutation in eukaryotes. Annu Rev Genet 5:163–218

Averbeck D (2010) Non-targeted effects as a paradigm breaking evidence. Mutat Res 687:7–12

Aypar U, Morgan WF, Baulch JE (2011) Radiation-induced epigenetic alterations after low and high LET irradiations. Mutat Res 707:24–33

Barber RC, Hickenbotham P, Hatch T et al (2006) Radiation-induced transgenerational alterations in genome stability and DNA damage. Oncogene 25:7336–7342

Belyakov OV, Folkard M, Mothersill CE et al. (2000) Bystander effect and genomic instability. Challenging the classic paradigm of radiobiology. In: Proceedings Timofeeff-Ressovsky centennial conference "Modern problems of Radiobiology, Radioecology and Evolution". JINR Press, Dubna

Belyakov OV, Hall EJ, Marino SA et al. (2002) Studies of bystander effects in artificial human 3d tissue systems using a microbeam irradiation. Annual report. Center for Radiological Research, Columbia University, Irvington

Bliznik KM, Kapultsevich YG, Korogodin VI et al (1974) Formation of radioraces by yeasts. Comm. 4. The dependence of the saltant yield on postirradiation cultivation conditions. Radiologiya 14:230–236 (Russian)

Boei JJ, Vermeulen S, Natarajan AT (1996) Detection of chromosomal aberrations by fluorescence in situ hybridization in the first three postirradiation divisions of human lymphocytes. Mutat Res 349:127–135

Brenner DJ, Little JB, Sachs RK (2001) The bystander effect in radiation oncogenesis: II. A quantitative model. Radiat Res 155:402–408

Brenner DJ, Sachs KD (2002) Do low dose-rate bystander effects influence domestic radon risks? Int J Radiat Biol 78:593–604

Burlakova EB (1994) Effect of the minute doses. Bull Russ Acad Sci 4:80–95 (Russian)

Burlakova EB, Goloschapov AN, Gorbunova NV et al (1996) Features of the low-dose-radiation biological effects. Radiats Biol Radioecol 36:610–631 (Russian)

Burlakova EB, Goloschapov AN, Zhizhina GP et al (1999) New view on the regularities of low-dose-rate irradiation at the low doses irradiation. Radiats Biol Radioecol 39:26–34 (Russian)

Burlakova EB, Mikhailov VF, Mazurik VK (2000) System of the oxidation-reduction homeostasis at radiation- inducible genome instability. Radiats Biol Radioecol 41:489–499 (Russian)

Cherezhanova LV, Alexakhin RM (1971) To the question of cytogenetic many-year influence of high artificial radiation on populations. Russ J Gen Biol 32:494–500 (Russian)

Deshpande A, Goodwin EH, Bailey SM et al (1996) α-particle-induced sister chromatid exchange in normal human lung fibroblasts: evidence for an extranuclear target. Radiat Res 145:260–267

Dubrova YE (2003) Radiation-induced transgenerational instability. Oncogene 22:7087–7093

Dubrova YE (2012) Genomic instability in the offspring of irradiated parents. In: Mothersill C, Korogodina V, Seymour C (eds) Radiobiology and environmental security. Springer, Dordrecht, pp 127–140

Dubrova YE, Plumb M, Brown J et al (2000) Induction of minisatellite mutations in the mouse germline by low-dose chronic exposure to γ-radiation and fission neutrons. Mutat Res 453:17–24

Ellegren H, Lindgren G, Primmer CR et al (1997) Fitness loss and germline mutations in barn swallows breeding in Chernobyl. Nature 389:593–596

Emerit I, Oganesian N, Arutyunian R et al (1997) Oxidative stress-related clastogenic factors in plasma from Chernobyl liquidators: protective effects of antioxidant plant phenols, vitamins and oligoelements. Mutat Res 377:239–246

Feinendegen LE (2005) Evidence for beneficial low level radiation effects and radiation hormesis. Br J Radiol 78:3–7

Florko BV, Korogodina VL (2007) Analysis of the distribution structure as exemplified by one cytogenetic problem. PEPAN Lett 4:331–338

Fomenko LA, Vasil'eva GV, Bezlepkin VG (2001) Micronucleus frequency is increased in bone marrow erythrocytes from offspring of male mice exposed to chronic low-dose gamma irradiation. Biol Bull 28:419–423

Geras'kin SA, Oudalova AA, Kim JK et al (1998) Analysis of cytogenetic effects of low dose chronic radiation in agricultural crops. Radiats Biol Radioecol 38:367–374 (Russian)

Goh K, Sumner H (1968) Breaks in normal human chromosomes: are they induced by a transferable substance in the plasma of persons exposed to total body irradiation? Radiat Res 35:171–181

Gudkov IN (1985) Cell mechanisms of postradiation repair in plants. Naukova Dumka, Kiev (Russian)

Hickman A, Jaramillo R, Lechner J et al (1994) Alpha-particle-induced p53 protein expression in a rat lung epithelial cell strain. Cancer Res 54:5797–5800

Jaenisch R, Bird A (2003) Epigenetic regulation of gene expression: how the genome integrates intrinsic and environmental signals. Nat Genet 33(Suppl):245–254

Jirtle RL, Skinner MK (2007) Environmental epigenomics and disease susceptibility. Nat Rev Genet 8:253–262

Kadhim MA, Hill MA, Moore SR (2006) Genomic instability and the role of radiation quality. Radiat Prot Dosimetry 122:221–227

Kapultsevich YG (1978) Quantitative regularities of cell radiation injury. Atomizdat, Moscow (Russian)

Kapultsevich YG, Petin VG (1977) Probability model for cell responses to irradiation. Studia Biophysika 62:151

Kashino G, Prise KM, Suzuki K et al (2007) Effective suppression of bystander effects by DMSO treatment of irradiated CHO cells. J Radiat Res (Tokyo) 48:327–333

Kaup S, Grandjean V, Mukherjee R et al (2006) Radiation-induced genomic instability is associated with DNA methylation changes in cultured human keratinocytes. Mutat Res 597:87–97

Korogodin VI (1993) The school of N. W. Timofeeff-Ressovsky. In: Vorontsov NN (Ed) Nikolay Wladimirovich Timofeeff-Ressovsky. Stories, recollections, materials. Nauka, Moscow, pp 252–269 (Russian)

Korogodin VI, Bliznik KM, Kapultsevich YG (1977) Regularities of radioraces formation in yeasts. Comm. 11. Facts and hypotheses. Radiologiya 17:492–499 (Russian)

Korogodin VI, Bliznik KM (1972) Formation of radioraces by yeasts. Comm. 1. Radioraces of diploid yeasts *Saccaromyces ellipsoideus vini*. Radiologiya 12:163–170 (Russian)

Korogodin VI, Bliznik KM, Kapultsevich YG et al (1972) Formation of radioraces by yeasts. Comm. 3. The quantitative regularities of radiorace formation by diploid yeasts. Radiologiya 12:857–863 (Russian)

Korogodina VL, Florko BV, Osipova LP (2010) Adaptation and radiation-induced chromosomal instability studied by statistical modeling. Open Evol J 4:12–22

Kovalchuk O, Baulch JE (2008) Epigenetic changes and nontargeted radiation effects – is there a link? Environ Mol Mutagen 49:16–25

Kovalchuk O, Dubrova YE, Arkhipov A et al (2000) Wheat mutation rate after Chernobyl. Nature 407:583–584

Kovalchuk O, Kovalchuk I, Arkhipov A et al (2003) Extremely complex pattern of microsatellite mutation in the germline of wheat exposed to the post-Chernobyl radioactive contamination. Mutat Res 525:93–101

Kuzin AM (1993) The key mechanisms of radiation hormesis. Izv Akad Nauk Ser Biol 6:824–832

Lacassagne A, Schoen M, Beraud P (1939) Contribution à l'étude des radio-races de levures. II. Caracté res physiologiques de quelques radio-races d'une levure de vin. Ann Fermentations 5:129–152 (French)

Lea DE (1946) Action of radiations on living cells. Cambridge University Press, Cambridge

Leonard BE (2008a) A review: development of a microdose model for analysis of adaptive response and bystander dose response behavior dose response. Int J Radiat Biol 6(2):113–183

Leonard BE (2008b) A composite microdose adaptive response (AR) and bystander effect (BE) model-application to low LET and high LET AR and BE data. Int J Radiat Biol 84:681–701

Liang X, So YH, Cui J et al (2011) The low-dose ionizing radiation stimulates cell proliferation via activation of the MAPK/ERK pathway in rat cultured mesenchymal stem cells. J Radiat Res (Tokyo) 52(3):380–386

Little JB (2000) Radiation carcinogenesis. Carcinogenesis 21:397–404

Little JB, Morgan WF (guest eds) (2003) Special issue on Genomic instability. Oncogene 13(22): 6977

Little JB, Nagasawa H, Pfenning T et al (1997) Radiation-induced genomic instability: delayed mutagenic and cytogenetic effects of X-rays and alpha particles. Radiat Res 148:299–307

Little MP (2010) Do non-targeted effects increase or decrease low dose risk in relation to the linear-non-threshold (LNT) model? Mutat Res 687:17–27

Liu Z, Mothersill CE, McNeill FE et al (2006) A dose threshold for a medium transfer bystander effect for a human skin cell line. Radiat Res 166(1 Pt 1):9–23

Longerich S, Galloway AM, Harris RS et al (1995) Adaptive mutation sequences reproduced by mismatch repair deficiency. Proc Natl Acad Sci USA 92:12017–12020

Lorimore SA, Coates PJ, Wright EG (2003) Radiation-induced genomic instability and bystander effects: inter-related nontargeted effects of exposure to ionizing radiation. Oncogene 22:7058–7069

Lorimore SA, Wright EG (2003) Radiation-induced genomic instability and bystander effects: related inflammatory-type responses to radiation-induced stress and injury? A review. Int J Radiat Biol 79:15–25

Luchnik NV (1958) Influence of low-dose irradiation on mitosis of pea. Bull MOIP Ural Department 1:37–49 (Russian)

Luckey TD (1980) Hormesis with ionizing radiation. CRC Press, Boca Raton

Luke GA, Riches AC, Bryant PE (1997) Genomic instability in haematopoietic cells of F_1 generation mice of irradiated male parents. Mutagenesis 12:147–152

Luning KG, Frolen H, Nilsson A (1976) Genetic effects of ^{239}Pu salt injections in male mice. Mutat Res 34:539–542

Lyng FM, Maguire P, McClean B et al (2006) The involvement of calcium and MAP kinase signaling pathways in the production of radiation induced bystander effects. Radiat Res 165:400–409

McClintock B (1938) The production of homozygous deficient tissues with mutant characteristics by means of the aberrant mitotic behavior of ring-shaped chromosomes. Genetics 23:315–376

McClintock B (1984) The significance of responses of the genome to challenge. Science 226:792–801

Morgan WF (2003a) Non-targeted and delayed effects of exposure to ionizing radiation: I. Radiation-induced genomic instability and bystander effects *in vitro*. Radiat Res 159:567–580

Morgan WF (2003b) Non-targeted and delayed effects of exposure to ionizing radiation. II. Radiation-induced genomic instability and bystander effects in vivo, clastogenic factors and transgenerational effects. Radiat Res 159:581–596

Morgan WF (2011) Radiation induced genomic instability. Health Phys 100:280–281

Mothersill CE, Seymour CB (1997) Medium from irradiated human epithelial cells but not human fibroblasts reduces the clonogenic survival of unirradiated cells. Int J Radiat Biol 71:421–427

Mothersill C, Seymour CB (1998) Cell–cell contact during γ-irradiation is not required to induce a bystander effect in normal human keratinocytes: evidence for release of a survival controlling signal into medium. Radiat Res 149:256–262

Mothersill CE, Seymour CB (2000) Genomic instability, bystander effect and radiation risks: implications for development of protection strategies for man and environment. Radiats Biol Radioecol 40:615–620

Mothersill CE, Seymour CB (2001) Radiation-induced bystander effects: past history and future directions. Radiat Res 155:759–767

Mothersill CE, Seymour CB (2002) Relevance of radiation-induced bystander effect for environmental risk assessment. Radiats Biol Radioecol 42:585–587

Mothersill CE, Seymour CB (2004) Radiation-induced bystander effects and adaptive responses – the Yin and Yang of low dose radiobiology? Mutat Res 568:121–128

Mothersill C, Seymour RJ, Seymour CB (2004) Bystander effects in repair-deficient cell lines. Radiat Res 161:256–263

Mothersill C, Seymour CB (2005) Radiation-induced bystander effects: are they good, bad or both? Med Confl Surviv 21:101–110

Mothersill C, Seymour CB (2006) Radiation-induced bystander effects and the DNA paradigm: an "out of field" perspective. Mutat Res 597(1–2):5–10

Nadson GA, Filippov GS (1925) Influence des rayons X sur la sexualite' et la formation des mutantes chez les champignons infe'rieurs (Mucoriné es). Compt Rend Soc Biol 93:473–475 (French)

Nadson GA, Filippov GS (1932) Formation of new resistant races of microorganisms under the action of X-rays. Radioraces of Sporobolomyces. Vestn Rentgenol Radiolog 10:275–299 (Russian)

Nagar S, Smith LE, Morgan WF (2003) Characterization of a novel epigenetic effect of ionizing radiation: the death-inducing effect. Cancer Res 63:324–328

Nagasawa H, Huo L, Little B (2003) Increased bystander mutagenic effect in double strand break repair-deficient mammalian cells. Int J Radiat Biol 79:35–41

Nagasawa H, Little JB (1992) Induction of sister chromatid exchanges by extremely low doses of α-particles. Cancer Res 52:6394–6396

Nikjoo H, Khvostunov IK (2004) A theoretical approach to the role and critical issues associated with bystander effect in risk estimation. Hum Exp Toxicol 23:81–86

Okada M, Okabe A, Uchihori Y et al (2007) Single extreme low dose/low dose rate irradiation causes alteration in lifespan and genome instability in primary human cells. Br J Cancer 96:1707–1710

Parsons WB, Watkins CH, Pease GL et al (1954) Changes in sternal bone marrow following roentgen-ray therapy to the spleen in chronic granulocytic leukaemia. Cancer 7:179–189

Petin VG, Morozov II, Kabakova NM et al (2003) Some features of radiation hormesis in bacteria and yeast cells. Radiats Biol Radioecol 43:176–178 (Russian)

Pfeiffer P (1998) The mutagenic potential of DNA double-strand break repair. Toxicol Lett 96–97:119–129

Portess DI, Bauer G, Hill MA et al (2007) Low-dose irradiation of nontransformed cells stimulates the selective removal of precancerous cells via intercellular induction of apoptosis. Cancer Res 67:1246–1253

Pozolotina VN (1996) Radiation-induced adaptation processes in plants. Ecology 2:111–116 (Russian)

Prise KM (2006) New advances in radiation biology. Occup Med (Lond) 56(3):156–161

Prise KM, Belyakov OV, Folkard M et al (1998) Studies of bystander effects in human fibroblasts using a charged particle microbeam. Int J Radiat Biol 74:793–798

Rigaud O, Moustacchi E (1996) Radioadaptation for gene mutation and the possible molecular mechanisms of the adaptive response. Mutat Res 358:127–134

Sawant SG, Randers-Pehrson G, Geard CR et al (2001) The bystander effect in radiation oncogenesis: I. Transformation in C3H 10 T1/2 cells in vitro can be initiated in the unirradiated neighbors of irradiated cells. Radiat Res 155:397–401

Schettino G, Folkard M, Prise KM et al (2003) Low-dose studies of bystander cell killing with targeted soft X-rays. Radiat Res 160:505–511

Schettino G, Folkard M, Michael BD et al (2005) Low-dose binary behavior of bystander cell killing after microbeam irradiation of a single cell with focused C(k) X-rays. Radiat Res 163:332–336

Sedelnikova OA, Nakamura A, Kovalchuk O et al (2007) DNA double-strand breaks form in bystander cells after microbeam irradiation of three-dimensional human tissue models. Cancer Res 67:4295–4302

Seymour CB, Mothersill CE (1997) Delayed expression of lethal mutations and genomic instability in the progeny of human epithelial cells that survived in a bystander killing environment. Radiat Oncol Investig 5:106–110

Shao C, Lyng FM, Folkard M et al (2006) Calcium fluxes modulate the radiation-induced bystander responses in targeted glioma and fibroblast cells. Radiat Res 166:479–487

Shao C, Folkard M, Prise KM (2008) Role of TGF-beta1 and nitric oxide in the bystander response of irradiated glioma cells. Oncogene 27:434–440

Shevchenko VA, Pechkurenkov VL, Abramov VI (1992) Radiation genetics of the native populations. Genetic consequences of the Kyshtym accident. Nauka, Moscow

Shevchenko VA, Kal'chenko VA, Abramov VI et al (1999) Genetic effects in populations of plants growing in the zone of Kyshtym and Chernobyl accidents. Radiats Biol Radioecol 39:162–176 (Russian)

Sigg M, Crompton NE, Burkart W (1997) Enhanced neoplastic transformation in an inhomogeneous radiation field: an effect of the presence of heavily damaged cells. Radiat Res 148:543–547

Timofeeff-Ressovsky NW (1939) Genetik und evolution. Z Inductive Abstammungs Vererbungslehre 76:158–218 (German)

Timofeeff-Ressovsky NW (1956) Biophysics interpretation of the radiostimulation phenomenon in plant. Biophysics 1:616–627 (Russian)

Timofeeff-Ressovsky NW, Poryadkova NA, Preobrazhenskaya EI (1950–1954) Influence of low dose irradiation on growth of plants. Reports. Fund of the Urals Department of the Academy of Sciences of USSR (Russian)

Timofeeff-Ressovsky NW, Timofeeff-Ressovskaya EA (2006) The principle types of planned experiments on radiation biogeocenology of freshwater communities. In: Korogodina VL, Cigna AA, Durante M (eds) Proceedings of the second international conference dedicated to NW Timofeeff-Ressovsky. JINR Press, Dubna, vol 2, pp 14–16 (Russian)

Timofeeff-Ressovsky NW, Tyuryukanov AN (2006) The principle types of planned experiments on radiation biogeocenology of terrestrial communities. In: Korogodina VL, Cigna AA, Durante M (eds) Proceedings of the second international conference dedicated to NW Timofeeff-Ressovsky. JINR Press, Dubna, vol 2, pp 17–19 (Russian)

Timofeeff-Ressovsky NW, Zimmer KG (1947) Biophysik. Das Trefferprinzip in der Biologie. S. Hirzel Verlag, Leipzig

Timofeeff-Ressovsky NW, Zimmer KG, Delbrück M (1935) Über die Nature der Genmutation und der Genstruktur. Nachr Ges Wiss Gottingen FG VI Biol NF 1:189–245

de Toledo SM, Buonanno M, Li M et al (2011) The impact of adaptive and non-targeted effects in the biological responses to low dose/low fluence ionizing-radiation: the modulating effect of linear energy transfer. Health Phys 100:290–292

Upton AC (2001) Radiation hormesis: data and interpretations. Crit Rev Toxicol 31(4–5):681–695

Vaiserman AM, Mekhova LV, Koshel NM et al (2010) Cancer incidence and mortality after low-dose radiation exposure: epidemiological aspects. Radiats Biol Radioecol 50:691–702

Vorobtsova IE (2000) Irradiation of male rats increases the chromosomal sensitivity of progeny to genotoxic agents. Mutagenesis 15:33–38

Vorobtsova IE, Aliyakparova LM, Anisimov VN (1993) Promotion of skin tumors by 12-O-tetradecanoylphorbol-13-acetate in two generations of descendants of male mice exposed to X-ray irradiation. Mutat Res 287(2):207–216

Watson GE, Lorimore SA, MacDonald DA et al (2000) Chromosomal instability in unirradiated cells induced *in vivo* by a bystander effect of ionizing radiation. Cancer Res 60:5608–5611

Wiley LM, Baulch JE, Raabe OG et al (1997) Impaired cell proliferation in mice that persists across at least two generations after paternal irradiation. Radiat Res 148:145–151

Zaichkina SI, Rozanova OM, Akhmadieva AK et al (2009) Study of the genetic instability in generations of mice irradiated of a low-dose rate of high-LET radiation. Radiats Biol Radioecol 49:55–59 (Russian)

Zhou H, Randers-Pehrson G, Waldren CA et al (2000) Induction of a bystander mutagenic effect of α-particles in mammalian cells. Proc Natl Acad Sci USA 97:2099–2104

Zyuzikov NA, Coates PJ, Parry JM et al (2011) Lack of nontargeted effects in murine bone marrow after low-dose *in vivo* X irradiation. Radiat Res 175(3):322–327

Chapter 3
Excursus on Statistical Modelling for Population Biology. Statistical Solution of Some Radiobiological Tasks

Abstract The consequences of background-level radiation often provoke debates, and here statistical ideas and their mathematical basis are considered in view of adaptation processes. Statistical modelling presents the data of investigations in the form of frequency function of events occurrence that allows studying the laws and regularities of respective probability processes. It is qualified to investigate the processes induced by the low factors and accompanied by Darwinian selection in different systems. Three themes are discussed: (1) A geometric model of adaptation, (2) Research of the biological communities' structure, and (3) A statistical view of the cytogenetic investigations of instabilities. The first and second topics present development of statistical ideas on live systems under environmental conditions. The last part is devoted to models of appearance of cells with abnormalities, chromosomal abnormalities in cells, proliferated cells and interrelation between the distributions on the number and frequency of abnormalities. We can assume that a strong factor leads to the same laws of regulated abundance of the communities of species and cell population.

Keywords Statistical modelling • adaptation process • geometric model of adaptation • biological communities' structure • appearance of cells with abnormalities • appearance of chromosomal abnormalities • proliferated cells occurrence • individual distribution on the frequency of abnormalities

3.1 Preliminaries

Let us consider the results of investigations in view of the processes of events occurrence but not mechanisms that produce these events. The statistical modelling allows us to separate each process and study its regularities.

The experimental data usually reflect a compound of the processes induced by different mechanisms, and the lab experimenters can exclude artificially one or two well-known interfering phenomena to analyze the necessary one. These laboratory methods are not suitable for ecology and epidemiology research. But all events follow their own statistical laws. Their statistical representation allows us to understand the nature of the phenomena, study their features, and calculate the risks. The statistical modelling helps us to analyze the combination of two or three processes in a way that is difficult through usual methods. An example of such a complex problem is the adaptation process induced by low-dose-rate irradiation comparable with a natural background; that is, the consequences of radiation fallout for plant populations growing within a 30-km zone of an operational nuclear station and the nuclear tests in the 1950s for human populations living in the polluted territories.

This chapter presents statistical ideas and a mathematical basis for the statistical modelling. It contains three sections: a geometric model of adaptation, research of the biological communities' structure, and a statistical view of the cytogenetic investigations of instabilities. The first and second sections discuss development of the statistical ideas on live systems under environmental conditions. These two sections consider investigations published by other researchers, and the last section corresponds to the mathematical studies of the adaptation process and genetic instability performed at the Joint Institute for Nuclear Research.

3.2 Geometric Model of Adaptation

Here, the statistical ideas are shown as to how adaptation can be represented. R.A. Fisher was the first to offer description of adaptation, and H.A. Orr showed that the long tails of statistical distributions suggested selection in the evolutionary model. These investigations are basic for a statistical representation of the adaptation processes.

Fisher's model: In "The Genetical Theory of Natural Selection" (1930) Ronald A. Fisher characterized adaptation as a movement of a population towards a phenotype according to environmental conditions, and formulated his so-called geometric model of adaptation. H.A. Orr investigated the Fisher model and specified its statements as follows: according to the Darwinian Theory, populations must adapt by using random mutations; mutations have different phenotypic 'sizes'; and populations must adapt in the face of pleiotropy (Orr 2005a).

In his model, Fisher put the optimal combination of the organisms' characteristics into the origin of the coordinate system. The environmental changes corresponded to a removal of population from this optimum, and a population must attempt to return by producing mutations. Fisher showed that the probability $P(x)$ appropriated to a given phenotypic size is $1 - F(x)$, where F is a function of a standard normal random variable x which is an inversely proportional quantity as distance to the optimum.

3.2 Geometric Model of Adaptation

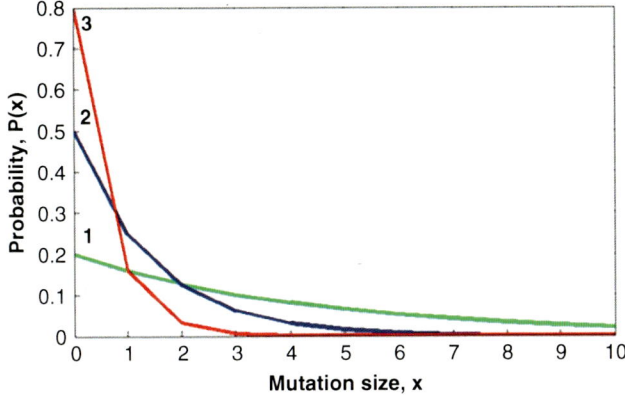

Fig. 3.1 Geometric model of adaptation. The x-axis is a mutational size, and the y-axis is a probability of random mutation of a given phenotypic size to be beneficial. 1, $P(x) = 0.2$; 2, $P(x) = 0.5$; 3, $P(x) = 0.8$

The probability of a mutation with the given phenotypic size being beneficial is plotted in Fig. 3.1. The plot shows that the probability falls rapidly with mutational size; very small mutations have a 20, 50 or 80 % chance of being favorable (see plots 1, 2 or 3, respectively) although the probability of the larger mutations falls with an increase of the mutational size. Fisher concluded that very small mutations should be considered as the genetic basis of adaptation. His model has suggested micromutationism.

But M. Kimura regarded this hypothesis as an erroneous one (1983). Kimura noted that micromutations could be lost when they are rare and mutations of a larger effect are more likely to escape such losses, therefore they could very well contribute to adaptation. Kimura concludes that mutations of intermediate sizes are the most beneficial.

Modern theory of adaptation: H.A. Orr used analytic methods as well as computer modelling and has shown that the size distribution of mutations is nearly exponential (Orr 1998, 1999). It means that adaptation involves a few mutations of a relatively large phenotypic effect and many of a relatively small effect, as Fisher's model predicts. These results revealed stability of the assumptions about the precise shape of the fitness function.

J.H. Gillespie performed fitness investigations using molecular evolutionary data and worked out the so-called mutational landscape model (Gillespie 1983, 1984). It is surprising that older phenotypic and modern DNA sequence-based models, in spite of their fundamental differences, agree with Fisher's geometric model (Orr 2005b).

Using mathematical methods, R.A. Fisher (1930), J.H. Gillespie (1983, 1984), and H.A. Orr (1998, 1999, 2006) investigated fitness to the environmental conditions by modelling. Orr (2006) showed a universal character of distributions with long tails and the connection of this phenomenon with selection in evolutionary models.

3.3 View on the Structure of Biological Communities

There are many points of view on the structure of biological communities which reflect different environmental conditions and dominant processes there (McGill et al. 2007). The scientists offered geometric, logarithmic, and lognormal laws to describe species abundance distribution. The outlook of I. Motomura (1932) was based on the limitation of environmental resources, which are, in fact, ideas of adaptation. Rich environmental conditions correspond to multiplication of the species (Preston 1948, 1962). Multiplication of the species can be presented as a process of fragmentation, which was described by A.N. Kolmogorov (1986). So, the proliferation of cells can be also described as a fragmentation process. Here, one example is considered when the biological community was transformed from the "rich" state to the "limited" one under radiation impact (the Totskij explosion (Vasiliev et al. 1997)).

Radiation geneticists used statistical modelling in radiation epidemiology to study the appearance of the multiaberrant cells in the blood of irradiated persons, and have revealed a combination of the geometric and Poisson distributions (Bochkov et al. 1972). This compound was also observed by R.M. Arutyunyan et al. (2001) who showed the appearance of multiple chromosomal aberrations in blood lymphocytes of patients with a syndrome of chromosomal instability.

These investigations have shown the statistical representation of adaptation processes in different live systems. These are quality studies, but they illustrate the possibilities of the statistical modelling.

3.3.1 Model of the Geometric Series of Motomura

I. Motomura (1932) assumed that the first dominant species captures quota k of some resources, the second most abundant species occupies the same quota k of the resources' excess, the third most abundant captures k of the excess and so on, until these limited resources are not divided between a total S species. If this condition is fulfilled, and the abundances of species (for example, their biomass or number of individuals) are proportional to their used quota of the resources, then the distribution of these abundances would be described by a geometric series (or a hypothesis of priority ecological niche occupation). In range/abundance coordinates, Motomura's distribution is presented by a straight line, the angle of elevation of which depends on the constant of geometric progression k. For the Motomura model, the design equation for S (number of species) and their abundance N_i ($i = 1, 2 \ldots S$) is

$$N_i = \frac{Nk(1-k)^{i-1}}{1-(1-k)^S},$$

where $N = \sum_{i=1}^{S} N_i$, k – parameters of the model.

3.3 View on the Structure of Biological Communities

The Motomura assumptions are not strong enough. Firstly, it is not explained to which resource it refers, because different types of resources limit the organisms of different species in any biological community. The second is that the capture intensity depends to a great extent on the nature of the species but not only on their quantity: for example, proportion of oxygen consumption can differ by thousands of degrees for different species.

3.3.2 Fisher's Log-Series

R.A. Fisher used an absolutely different approach to solve the problem of the community species structure. He worked at Rothamsted Experimental Station in England and helped entomologists A. Corbet and K. Williams to analyze the results of the night moths' capture. Fisher assumes that the distribution of the numbers of the moths species, which possessed a different quantity, was the best of all described by log-series (Fisher et al. 1943):

$$S = a(x + x^2/2 + x^3/3 + \cdots) = -a \ln(1 - x),$$

where a and x – parameters (indices of diversity); $a = N(1 - x)/x$.

According to Fisher's model, the maximal number of species corresponds to the category of very rare ones which are presented by only one copy. Log-series are characterized by a small number of abundant species and a great quota of "rare" ones. The communities, whose structure is determined by one or a few ecological factors, can be with a high probability described by such a log-series.

3.3.3 Preston's Lognormal Distribution of Species Abundances

A little bit later, investigations by F.W. Preston were published (1948). Preston used logarithmic classes of abundance, which he called "octaves". The first octave consists of one individual; the second one consists of two individuals, and so on. A histogram of such distribution resembles the Gauss curve of the normal distribution. In other words, the distribution was lognormal: very rare species were observed more seldom than slightly more numerous species. Preston supposed that all sufficiently large samples of the community should be described by a lognormal distribution. If a sample is insufficiently representative, then the left curve portion of the normal distribution (the "tail" of rare species) should be cut-off.

B.J. McGill et al. (2007) published an analysis of more than 40 different models of species abundance distributions including well-known investigations by K.O. Winemiller (1990), R.H. Whittaker (1960), R. Condit et al. (1996), C.S. Robbins et al. (1986), and many others. In spite of this great number of different models, all the models demonstrated satisfactorily a "hollow curve" (that is, the distribution characterized by a great number of rare species and a very small number of mass species).

Enormous discussions were taking place on the nature of the left side of species abundance distributions when plotted on a log-abundance scale. Preston explained this with the concept of a veil line (Preston 1948). R.H. Whittaker (1965) attempted to resolve this debate by suggesting that no one curve fits all the data – the geometric model applied to species poor communities while the lognormal applied to more species rich communities, only being fully unveiled in large samples.

3.3.4 Studies of the Biological Communities That Experienced the Totskij Nuclear Explosion

It is interesting to consider the extensive investigations of the consequences of the Totskij nuclear explosion (1954, Orenburg region, Russia) provided by authors of the book *"Ecogenetic Analysis of Late Consequences of the Totskij Nuclear explosion in Orenburg Region* in 1954 (facts, models, hypotheses)" (Vasiliev et al. 1997). The authors investigated the environment by radiochemical, radioecological, cytogenetic, phenogenetic, and medicinal methods. Their investigations demonstrated significant elevation of the chromosomal aberrations and morphologic abnormalities of the mutational nature in species of small mammals. The numbers of species in the polluted and control sites were studied for a community of small mammals. These data indicated the degradation of the impact environment compared with the control one. In the control, the curve of the species numbers agrees with the lognormal distribution of F.W. Preston whereas in the impact sites it corresponds to the geometric series of I. Motomura (Vasiliev et al. 1997). The authors conclude that these results indirectly show that success of the coexistence of species in impact sites is determined by a bigger number of limiting factors in comparison with the control territory.

3.3.5 Influence of the Mutagenic Factor on Cellular Population

On the basis of the hit and target theory (see Chap. 2), D.E. Lea (1946) suggested that the distribution of cells on the number of aberrations induced by irradiation must follow the Poisson law. N.P. Bochkov and his colleagues investigated human cells after chemical exposure *in vivo* and *in vitro*. They assumed that the action of chemical mutagens on cells could be described by queuing theory (Bochkov et al. 1972), and for this reason a geometric distribution of cells on the number of aberrations can be expected. The experimental data and fitting obtained by Bochkov and his co-authors (Bochkov et al. 1972), is presented in Fig. 3.2.

The geometric law prevails under Poisson at any dose. The authors offered the following argumentation to clarify their results. When mutagenic molecules come into a cell independently and during the fixed time, the number of these molecules can be approximated by the Poisson law:

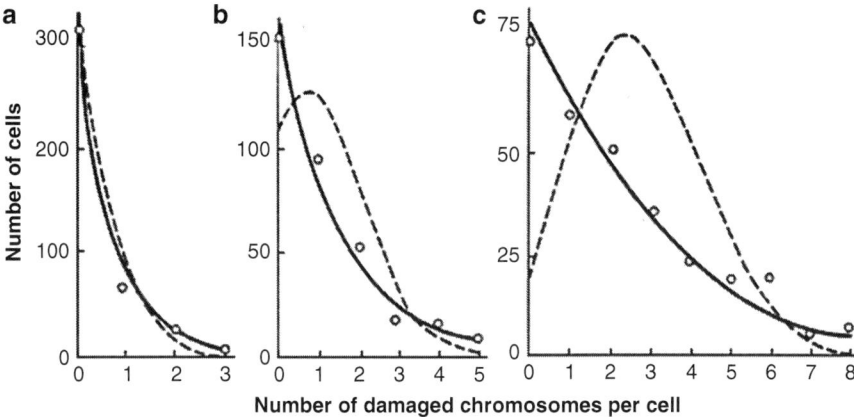

Fig. 3.2 The comparison of experimental data (*circles*) with Poisson (*the dotted line*) and geometric (*the solid line*) distributions. Cells were treated with thiophosphamide: **a, b, c** – 10, 20, and 30 γ/ml, respectively. x-axis – number of damaged chromosomes (classes) per cell; y-axis – number of cells (Figure is republished from (Bochkov et al. 1972))

$$P_n(t) = \frac{(\lambda t)^n e^{-\lambda t}}{n!},$$

where λ – is the average intensity of the mutagenic molecule reception.

The cell inactivates the mutagenic molecules at their reception; the aberrations correspond to the cases when the cell hasn't time to inactivate the receipted mutagen. Let us take into account the cell intensity of mutagen inactivation (μ), then $\rho = \lambda/\mu$ is the level of the aberrant cells. In this case, the cells are geometrically distributed on the number of aberrations.

N.A. Chebotarev offered the following model of the multiaberrant cell appearance (Chebotarev 2000). He postulates that the total cells' population is divided into two subpopulations. The first subpopulation consists of the cells where chromosomal aberrations appear independently due to external factors, and such cells are Poisson-distributed on the number of aberrations (P). The second subpopulation is characterized by the "infection factor", which promotes the aberrations' appearance in the cells under the same external mutagenic factors. In this second subpopulation, the cells are distributed on the aberrations' number according to the geometric law (G). The model can be described by the following system of equations:

$$T = P + G$$

$$T_0 = P_0 + G_0 = Pe^{-m} + G\theta$$

$$T_1 = P_1 + G_1 = Pe^{-m}m + G\theta(1 - \theta)$$

$$T_2 = P_2 + G_2 = Pe^{-m}m^2/2 + G\theta(1 - \theta)^2$$

where T – is the total number of cells, T_0 – the total number of cells without aberrations, T_1 – the total number of cells with one aberration, and T_2 – the total number of cells with two aberrations. P_0, P_1, and P_2 are the same, but for the Poisson subpopulation, and G_0, G_1, and G_2 for the geometric one; m is the sample mean of the Poisson distribution, and θ is the parameter of the geometric distribution. This model describes well the published data (Bochkov and Chebotarev 1989) on the appearance of multiple mutations in lymphocytes of individuals living in the town of Seversk who worked for the town's radiochemical industry (Chebotarev 2000).

3.4 Statistical View on the Cytogenetic Investigations of Instabilities

Radiation is the factor which leads to the processes of adaptation. Radiation stress induces the instability and selection processes which result in the long tails of distributions on the number of abnormalities (Korogodina et al. 2010a). The statistical modelling allows one to study the main inter- and intracellular processes induced by radiation stress. Some of these investigations were performed together with professors V.B. Priezzhev and G.A. Ososkov at the Joint Institute for Nuclear Research.

The period of plant seed germination was chosen for analysis of the adaptation mechanisms. The appearance of proliferated and abnormal cells in the seedling meristem is random and their numbers depend on exogenous and endogenous factors. Our task was to study the structure of seed distribution with respect to the number of abnormal (aberrant) and proliferated cells in the rootlets' meristem of the seeds which experienced low-radiation impacts.

The forming and features of the individuals' distribution on the frequency of cells with abnormalities in blood lymphocytes are considered at the end of this chapter.

3.4.1 Hypotheses and Models of Appearance of Cells with Abnormalities in Seedlings Meristem of Stress-Irradiated Seeds[1]

Let us assume that the number m of cells with chromosomal abnormalities (CCAs) in the meristem of a plant is described by the Poisson distribution:

$$P_m = \frac{\lambda^m}{m!} e^{-\lambda}$$

where λ is a mathematical expectation of m, or mean of the distribution.

---

[1] The full paper is published in (Florko and Korogodina 2007).

3.4 Statistical View on the Cytogenetic Investigations of Instabilities

We assume that plants in the population differ by the set of features denoted by one vector variable u. Parameters u, individual for each plant, can change parameter λ. Let us find a probability that seeds with m cells having chromosomal abnormalities (CAs) are found in the seed population. Assume that the fraction of plants with different parameters u and similar λ is equal to $p(\lambda)$. The probability that the seed belongs to the part of the population where the cells with CAs distribution are described by the Poisson distribution with the parameter in the interval $(\lambda, \lambda + d\lambda)$, is equal to $p(\lambda)d\lambda$. Then the probability of observing in this plant subpopulation the meristem with the number m of CCAs is $dP_m = \frac{\lambda^m}{m!} e^{-\lambda} p(\lambda) d\lambda$. The probability of observing in the whole population the meristems with number m of the CCAs is described by the Mandel formula:

$$P_m = \int_0^\infty dP_m = \int_0^\infty \frac{\lambda^m}{m!} e^{-\lambda} p(\lambda) d\lambda.$$

The theory of the inverse Mandel transformation has been developed, i.e., the method of reconstruction of $p(\lambda)$ from the given distribution P_m is known (Klauder and Sudarshan 1968). For this purpose, the following function should be constructed:

$$Q(x) = \sum_{m=0}^\infty (1-x)^m P_m = \int_0^\infty e^{-x\lambda} p(\lambda) d\lambda.$$

Then the function $R(x) = Q(ix) \int_0^\infty e^{-ix\lambda} p(\lambda) d\lambda$ is a characteristic function of the distribution $p(\lambda)$ (Gnedenko 1965). This function has a number of properties. One of them is important for us:

$$|R(x)| \leq R(0). \tag{3.1}$$

This property provides a possibility to verify the validity of the hypothesis of a similar cell reaction to the radiation impact. If the distribution P_m of plants with respect to the number of the CCAs is known, the function $R(x)$ can be constructed, and satisfaction of inequality (3.1) is checked. The verification carried out for our experimental data[2] and seed distributions with respect to the CCAs number has shown that in all cases, inequality (3.1) is violated. The typical plot of $R(x)$ is shown in Fig. 3.3. It is clear that $R(x)$ can exceed $R(0) = 1$, therefore we have to reject the hypothesis of an independent and similar cell reaction to the radiation impact.

Let us consider another hypothesis, according to which the cell reaction to irradiation is not independent. This violates the Poisson law of the CCAs occurrence in the meristem and results in other distributions. It is known that low dose radiation induces the bystander effect (Lorimore and Wright 2003), which provides the late

---

[2]Experimental data and approximations are presented in Chaps. 4, 5, and 8 (Sects. 8.2 and 8.4).

Fig. 3.3 Characteristic function of the distribution $R(x)$. x-axis – x; y-axis – $R(x)$

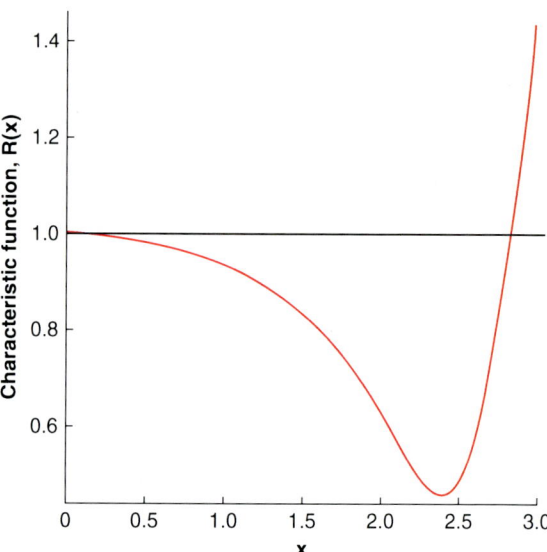

elevation of the abnormal cells number due to intercellular communication that results in the natural selection of seedlings. Then, the following assumptions can be made:

(i) Primary cell damage is described by Poisson statistics as the statistics of rare and independent events:

$$P_n = \frac{a^n}{n!} e^{-a},$$

where P_n is a probability of occurrence of n CCAs and a is the parameter of the Poisson distribution of the primary damages.

(ii) In some cell subpopulations of the meristem, intercellular communication processes can occur leading to new CCAs. This provides a variation of parameter a of the Poisson distribution. For small impacts, we assume that parameter a varies in time linearly, $a(t) = \lambda t + a$, and obtain the following formula:

$$P_n(t) = \frac{(\lambda t + a)^n}{n!} e^{-(\lambda t + a)},$$

where $P_n(t)$ is the probability of occurrence of n cells with CAs by time t, λ is the drift rate, and $a(t) = \lambda t + a$ is the parameter of the Poisson distribution at moment t.

(iii) Occurrence of the new CCAs increases the material for natural selection. We describe natural selection by the Markovian process in the phase space of

3.4 Statistical View on the Cytogenetic Investigations of Instabilities

the system parameters under the assumption of its stationary character (van Kampen 2007). The external conditions separate the region of the phase space, falling in which means adaptation of the meristem and termination of the process of the CCAs occurrence and selection.

These assumptions have analytic conclusions. It follows from general the theorems concerning Markovian processes (Harris 2002) that the probability of adaptation of the meristem at moment t is expressed by

$$G(t) = 1 - \sum_{i=1} \alpha_i e^{-\mu_i t},$$

where $\alpha_I > 0$ and $\mu_I > 0$ are sets of the parameters depending on characteristics of the processes of the CCAs occurrence and sprout selection. The probability Q_n of the observation of n CCAs in the meristem can be determined using the following transformations. The generating functions of the Poisson process $P(t)$ and probabilities Q_n are determined by the following:

$$P(t) = \sum_{i=0} P_{i(t)} z^i = e^{(\lambda t + a)(z-1)},$$

$$Q(z) = \sum_{i=0} Q_i z^i = \int_0^\infty P(t) dG(t)$$

$$= \int_0^\infty e^{(\lambda t + a)(z-1)} \left(\sum_{i=0} \alpha_i \mu_i e^{-\mu_i t} \right)$$

$$= e^{a(z-1)} \sum_{i=0} \alpha_i \mu_i \int_0^\infty e^{(\lambda(z-1) - \mu_i)t} dt$$

$$= e^{a(z-1)} \sum_{i=0} \alpha_i \mu_i \frac{1}{\mu_i - \lambda(z-1)}$$

$$= \sum_{i=0} \alpha_i \frac{\mu_i}{\mu_i + \lambda} \frac{e^{a(z-1)}}{1 - \frac{\lambda}{\mu_i + \lambda} z}$$

The generating function $Q(z)$ is a sum of the products of generating functions of the Poisson $e^{a(z-1)}$ and geometric distributions $\frac{\mu_i}{\mu_i + \lambda} \frac{e^{a(z-1)}}{1 - \frac{\lambda}{\mu_i + \lambda} z}$. If the average number of primary damaged cells of the subpopulation is small, $(a \approx 0)$, then $e^{a(z-1)} \approx 1$ and Q_n is described by the sum of geometric distributions. For $\alpha_k \gg \alpha_I$ and $i \neq k$, Q_n is described by the geometric distribution. The parameter $\frac{\lambda}{\lambda + \mu} = \frac{1}{1 + \mu/\lambda}$ of the geometric distribution depends on the ratio of rates of communication processes and selection process μ/λ. This ratio can be determined by the slope angle of the geometric distribution constructed on the semilogarithmic scale. The larger the slope angle, the more intensive the selection (as compared to the intercellular communications).

In fact, the population consists of resistant and sensitive subpopulations. The late cell damages can be accumulated without selection in the resistant fraction of seedlings because they are rare. The probabilities of communication processes and selection are higher in the sensitive fraction. The geometric law relates to the series of events which break under some conditions ("successful finish"), whereas the Poisson law corresponds to the numbers of rare and independent events (Feller 1957). In our case, the geometric law is determined by the successful selection of sensitive seedlings caused by the high intensity of late damages. Accumulation of rare damages without selection will be described by the Poisson law. Thus, the seed distribution with respect to the number of the CCAs in the seedlings meristem can be described by the sum of geometric and Poisson distributions ($G + P$). Their relation depends on the primary radiation intensity.

3.4.2 Correlative Model of Multiplication of the DNA Damages

In the meristem cells, the processes are like the above-mentioned ones: the primary and late damaging, and selection. The difference is that the multiple appearances of cell damages caused by the bystander effect in tissue are replaced by the radiation-induced genomic instability in cells (Morgan 2003).

Our model reasons are based on the following assumptions:

The main processes: All processes related to the DNA damages are assumed to be Markovian birth-death processes characterized by their rate constants a, b, and c (intensities) (Feller 1957). It means that the probability of the events appearance doesn't depend on the process history.

The processes which take place in the cell during its cycle are as follows:

1. appearance of independent DNA damages;
2. appearance of correlated DNA damages[3];
3. repair of the DNA damages.

Appearance of independent damages: In a cell with n damages, new damages can appear in any non-damaged targets ($N - n$), each of them transformed into a primary damaged target with intensity α.

The suggested mechanism of the damages' appearance means that intensity of the appearance is as follows:

$$a_n = \alpha (N - n),$$

---

[3] We suggested that the probability of the appearance of late damage in the non-damaged target due to induction by two or more damages is small.

3.4 Statistical View on the Cytogenetic Investigations of Instabilities

where n – is the number of damages; N – is the number of targets; α – is the rate of the appearance of primary damage in one target; a_n – is the rate of the appearance of primary damage in $(N-n)$ non-damaged targets.

Let us consider the first process as the birth-death process. The process is characterized by the system of the following differential equations (Feller 1957):

$$\frac{dP_n(t)}{dt} = a_{n-1} P_{n-1}(t) - a_n P_n(t) = \alpha(N - n + 1) P_{n-1}(t) - \alpha(N - n) P_n(t)$$

$$\frac{dP_0(t)}{dt} = -\alpha N P_0(t).$$

Appearance of correlated damages: The intensity of the appearance of late damages is proportional to n:

$$b_n = \lambda n,$$

where λ – is the intensity of the appearance of late damages; n – is the number of damages; b_n – is the rate of the appearance of late damages, which depends on the quantity of the existing damages n.

The appearance of late correlated damages can be described as the birth-death process by a system of differential equations:

$$\frac{dP_n(t)}{dt} = b_{n-1} P_{n-1}(t) - b_n P_n(t) = \lambda(n-1) P_{n-1}(t) - \lambda n P_n(t)$$

$$\frac{dP_1(t)}{dt} = -\lambda P_1(t).$$

Repair process: Let us accept a hypothesis that the rate of cell recovery is proportional to the quantity of the damages. It is equivalent to the proposition that the cells recover as a whole by the principle "either all or none" (Luchnik 1968).

Then, the repair mechanisms eliminate the damages with intensity

$$c_n = \mu n,$$

where μ – is the repair rate of one damage; n – is the number of damages; c_n – is the repair rate of the damage, which depends on the quantity of the existing damages n.

The repair process can be considered as the birth-death process, which will be described by a system of differential equations:

$$\frac{dP_n(t)}{dt} = -c_n P_n(t) + c_{n+1} P_{n+1}(t) = -cn P_n(t) + c(n+1) P_{n+1}(t)$$

$$\frac{dP_0(t)}{dt} = \mu P_1(t).$$

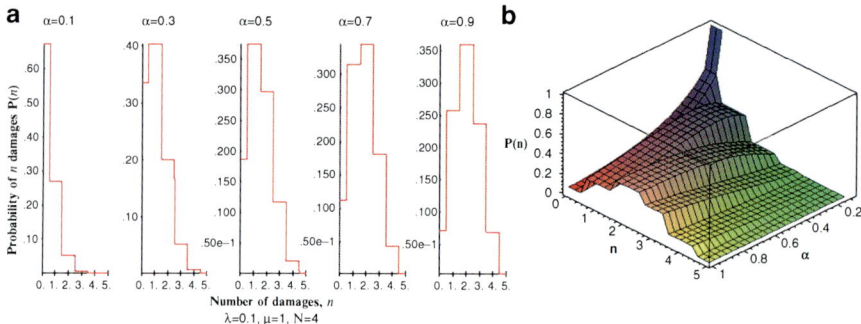

Fig. 3.4 Dependence of the probability of n damages $P(n)$ on the number of damages, n, for different α

The statistical model accounted for all of the above three processes: Three processes can be described by a system of the following equations:

$$\frac{dP_n(t)}{dt} = (a_{n-1} + b_{n-1})P_{n-1}(t) - (a_n + b_n + c_n)P_n(t) + c_{n+1}P_{n+1}(t)$$

$$\frac{dP_0(t)}{dt} = c_1 P_1(t), \qquad (3.2)$$

where $a_n + b_n = (\lambda - \alpha)n + N\alpha$; $c_n = \mu n$; n – is the number of damages; N – is the number of targets; α – is the intensity of the appearance of primary damage; λ – is the intensity of induction of correlated late damages; μ – is the intensity of the repair process.

The consequences of the correlative model: How do the intensities of processes α, μ and λ change the distribution of the number of damages?

The asymptotical solutions of (3.2) for $t \to \infty$ and various dependencies of α, μ and λ values are presented further in Figs. 3.4, 3.5, 3.6, 3.7, and 3.8.

Dependence on α (μ, λ = const):

Parameter α depends on dose rate irradiation, and n reflects the dose irradiation. At small α, the frequency of the damages has an exponential character (Fig. 3.4, $\alpha = 0.1$). The frequency of damages obtains the Poisson character with increasing of the α value (Fig. 3.4, $\alpha = 0.9$). The average number of damages per cell increases with the intensity of damaging α (Fig. 3.5).

Dependence on λ (α, μ = const):

With the increasing of the λ value (the parameter that describes late damage appearance) the character of the distribution doesn't change (Fig. 3.6), but the mean value of the distribution $<n>$ obtains major values according to the following formula:

$$<n> = \frac{N\alpha}{\alpha + \mu - \lambda} \qquad (3.3)$$

3.4 Statistical View on the Cytogenetic Investigations of Instabilities

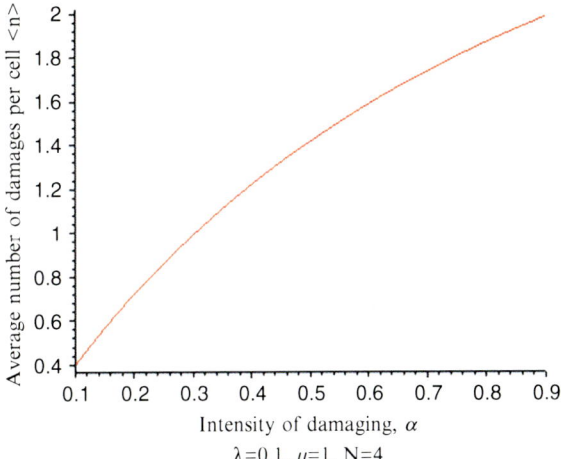

Fig. 3.5 The dependence of the average number of damages per cell on the intensity of damage α

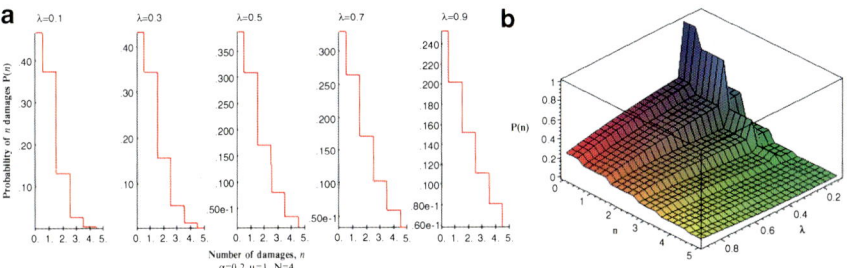

Fig. 3.6 Dependence of the probability of n damages $P(n)$ on the number of damages, n, for different λ

Dependence on μ (α, $\lambda = const$):

With the increasing of μ value (the parameter, which reflects repair of the damages), the character of the distribution doesn't change (the character of the distribution is determined by the α value) (Fig. 3.7); the mean $<n>$ of the distribution gets minor values according to the formula (3.3).

Independently of repair, the geometric distribution is observed in the control or very close to the control (Fig. 3.7).

At small μ value, the Poisson distribution is observed, whereas at the large one the null class increases, but the type of distribution remains (Fig. 3.8).

Why is Poisson distribution observed at the increasing of dose rate irradiation?

1. The appearance of the late correlative events is distributed in time.
2. The primary damages frequency increases in the cell while increasing the intensity (dose rate irradiation).

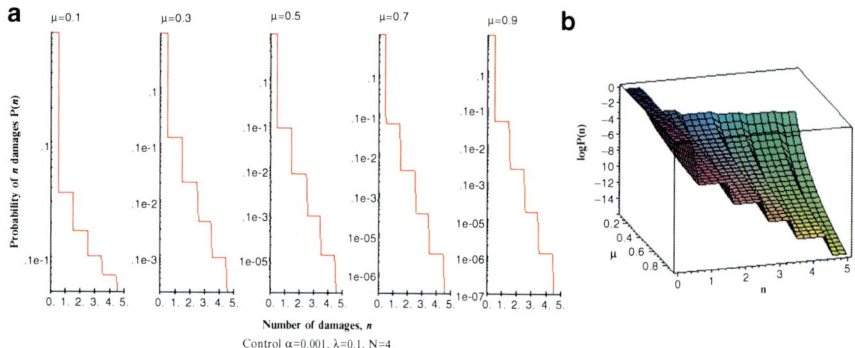

Fig. 3.7 Dependence of probability of n damages $P(n)$ on the number of damages, n, for different μ

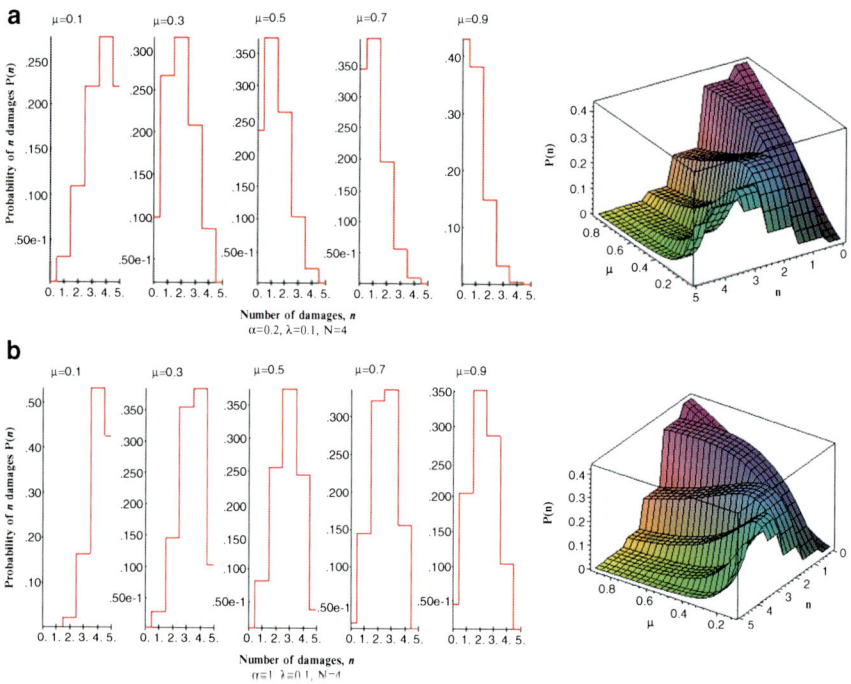

Fig. 3.8 Dependence of the probability of n damages $P(n)$ on the number of damages n for different μ. Low intensity of damaging $\alpha = 0.02$ (*top*); high intensity of damaging $\alpha = 1$ (*bottom*)

3. If the rate of the primary damage appearance exceeds the generation rate of tails of the geometric distribution (they are different), then those primary damages appearing earlier should generate the Poisson distribution according to queuing theory (Feller 1957). In addition to that, the quota of the geometric distribution will decrease.

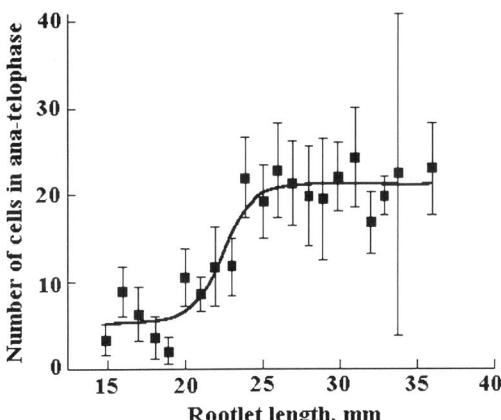

Fig. 3.9 Number of proliferated cells in the meristem of sprout rootlet as a function of the rootlet length

4. The cell can have a definite number of damages. Increasing the sum of the primary Poisson-like and late geometric-like damages can lead to the cells' death as a result of a large number of damages. At that time, the mean value of the geometric distribution will decrease.
5. Finally, the Poisson distribution should be observed.
6. Induction of the second repair system leads to the second geometric distribution, which describes the second group of cells. Its parameters depend on the dose rate irradiation.

These conclusions allow us to use the geometric and Poisson distributions to describe the appearance of chromosomal abnormalities in cells.

3.4.3 Hypothesis and Model of the Proliferated Cells Occurrence

Irradiation influences the number of proliferated cells, which can increase and decrease due to activation of the resting cells and the death of damaged ones. Here, we consider the statistical representation of the proliferated cells occurrence, the statistical description of the resting cells stimulation, and how radiation stress can change the seed distributions on the number of proliferated cells.

At the germination of seeds, the number of divided cells in rootlet meristem reaches a specific number, the average value of which is constant for the given environmental conditions and doesn't change later. Our method fixes this stationary level of the proliferated cells number at the stage of the first mitoses (Fig. 3.9) (Korogodina et al. 1998). The number of cells N_k at the stationary growth phase (at the stage of the first mitoses) can be described in terms of the stationary random branching process (Harris 2002), where k is the number of cell generation.

In this case, $K_k = N_k + 1/N_k$ are independent uniformly distributed random numbers. Obviously, $N_n = KN_1$, where $K = K_1 \times \cdots \times K_n$ and so $\log K = \log K_1 + \cdots + \log K_n$. The central limit theorem states: "If $X_1, \ldots, X_n$ are independent uniformly distributed random quantities with the statistical expectation and variance, then for $n \to \infty$ the distribution law of the sum $\sum_i X_i$ asymptotically approaches the normal distribution". In our case, $X_i = \log K_i$. It suggests that $\log K$ is distributed according to the normal law, and so K, and therefore N_n, are distributed according to the lognormal law (Kolmogorov 1986).[4]

The meristem is heterogeneous and the pool of proliferated cells consists of at least two subpopulations. Low-radiation doses additionally stimulate the resting cells to proliferation (Luchnik 1958). It results from the above that the hypothetic model of proliferated cells (PCs) occurrence can be represented as follows:

(i) At the stationary phase of the rootlet growth, the distribution of seed sprouts in the PC number is lognormal.[5]
(ii) PCs can occur in three independent subpopulations: two subpopulations reflecting the heterogeneous character of the proliferated pool and the third subpopulation corresponding to the resting cells stimulated to proliferation.
(iii) The lognormal law can be observed in the negligible environmental factors. The increasing of one of them – for example, radiation – leads to mutagenesis, selection, and the geometrical law.

It can be assumed that, in the general case, the number of PCs in the meristem is described by the sum of lognormal distributions,

$$PC_n = \sum_{i=1,k} \frac{A_i}{\sqrt{2\pi w_i} x} \exp\left(-\frac{(\ln \frac{x}{a_i})^2}{2w_i^2}\right),$$

where PC_n is the number of seeds with n PCs in the meristem of the plant, A_i is the value of the ith subpopulation, and a_i and w_i are the parameters of the lognormal distribution, $k = 1, 2, 3$. The alternative hypotheses were a single-component lognormal distribution (LN) and a combination of two lognormal distributions (2LN). In the case of the stress impact, one of the lognormal distributions on the PC numbers transforms into the geometric one (Florko and Korogodina 2007).

A sample of such a distribution of rootlets on the number of PCs is presented in Fig. 3.10.

---

[4] A.N. Kolmogorov studied the lognormal distribution of particle sizes under fragmentation. This process has to fulfill the condition that the sum of the fragment sizes does not exceed the whole particle size. This condition is fulfilled for any processes of reproduction in nature, as for cells, and for any live species.

[5] It is appropriate here to refer to F.W. Preston, who introduced a lognormal distribution into biology (Preston 1948, 1962). He described a distribution of the species in the natural community on their numbers by the lognormal law. The comparison inevitably comes to mind between the numbers of cells in sprout meristem and the numbers of species in the community. Maximal numbers of plants and species correspond to the fitness conditions for propagation of each of these objects.

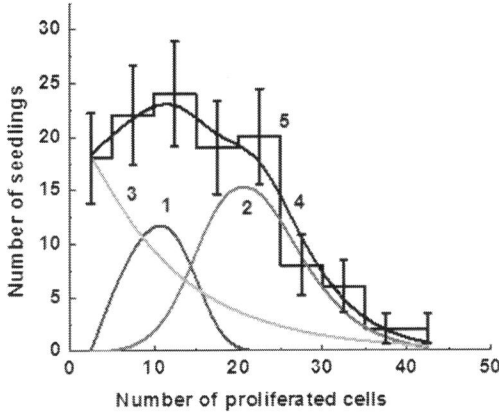

Fig. 3.10 Rootlet distribution with respect to the PC number in the population (Florko and Korogodina 2007). The lognormal (*1*, *2*) and geometric (*3*) components of the model distribution, their sum (*4*) are presented. Experimental distribution: the plot (*5*) with standard errors

The resting cells have elevated the number of mutations (Bridges 1997), and the death risk is high for the seedlings in the cell-activated subpopulation. Therefore, the resting cells stimulation has two evolutionary significances: it increases the pool of PCs and provides the material for natural selection.

3.4.4 Some Features of Forming and Analysis of Individuals' Distributions on the Number and Frequency of Cells with Abnormalities in Blood Lymphocytes

Radiation influences the lymphocyte cells of persons living in polluted areas, and the occurrence frequency of cells with abnormalities should be studied in the blood samples of individuals. The individuals' distributions on the number and frequency of cells with abnormalities are based on the law of appearance of cells with abnormalities among PCs, where number n of the PCs is a parameter. Therefore, individuals' distributions on the number and frequency of CCAs depend on individuals' distribution on the PC number.

Formally, this task is similar to the problem of studying the spectrum investigation observed by means of a non-ideal instrument (Kosarev 2008). The difference is that the distribution of CCAs is concentrated on integer numbers but not the real ones. Therefore, we have used an apparatus of the probability-generating functions instead of the Fourier transformation. We don't consider the processes of stimulation of proliferation and the death of cells.

Hypothesis of the appearance of PCs among the blood lymphocytes: PCs appear according to the discrete lognormal (LN) law (Florko et al. 2009). It means that a probability of appearance of n PC_n is described by

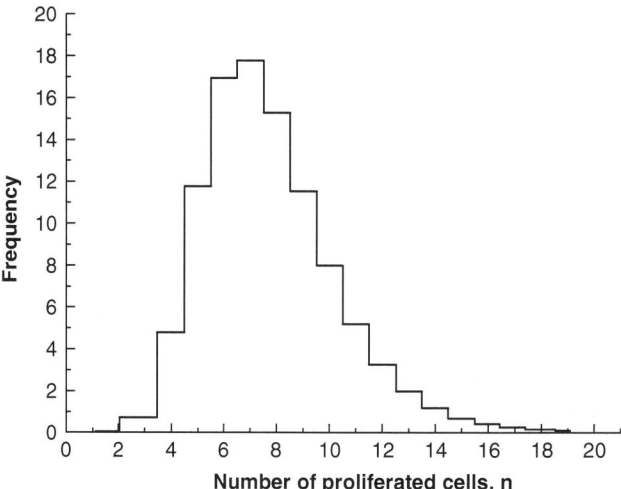

Fig. 3.11 Graph of lognormal distribution (modelling). Parameter values: $\mu = 2$, $\sigma = 0.1$; $m = 7.4$; $d = 0.74$

$$PC_n = \frac{1}{n\sigma\sqrt{2\pi}} e^{-(\ln n - \ln \mu)^2/(2\sigma^2)}$$

where μ, σ are distribution parameters.

For LN distribution (Fig. 3.11) the expectation m and dispersion d can be represented as distribution parameters by formulas $m = e^{\mu + \sigma^2/2}$ and $d = m\sqrt{e^{\sigma^2} - 1}$.

Hypothesis of appearance of CCAs among PCs in blood lymphocytes: We assume that spontaneous cell damages appear independently. Statistics of spontaneous events are binomial (Feller 1957). Low radiation induces the appearance of the number of primary cells with CAs that leads to time-dependent late processes of cells damaging (Lorimore and Wright 2003). The late stochastic processes terminate at a random moment, which depends on the affecting factor and radiosensitivity of the individual.

We have considered a time-dependent linear increase of the number of damaged cells and exponential distribution of termination moments (Florko et al. 2009) that leads to a special case of Pareto distribution (Reed and Hughes 2002), namely, to the exponential distribution on the number of CCAs. Other investigations have considered the case of an exponential increase in the number of CCAs with time, and the termination moments of this process are also exponentially distributed (Reed and Hughes 2002). It has been shown that the combined effects of the increase of the damage number and termination of this process lead to the power Pareto distributions (Florko et al. 2009).

3.4 Statistical View on the Cytogenetic Investigations of Instabilities

In the case of the binomial law, a probability of $CCA_{n,k}$ appearance of k CCAs among n PCs is described by formula $CCA_{n,k} = C_n^k q^k (1-q)^{n-k}$; in the geometric case, this probability can be described by formula $CCA_{n,k} = \frac{1-q}{1-q^n} q^k$, where q is a probability of the appearance of CCAs among the PCs.

Let us consider the cases of binomial and geometric laws of the appearance of CCAs among the PCs.[6]

Distribution of individuals on the number of CCAs in blood lymphocytes: Distribution of individuals on the number of CCAs (k) in blood lymphocytes can be described by the following formula:

$$CCA_k = \sum_{n=1..\infty} PC_n * CCA_{n,k},$$

where n is the number of PCs, and k is the number of CCAs.

Let us introduce the probability-generating function of the PC appearance

$$P(t) = \sum_{n=1..\infty} PC_n t^n$$

and the probability-generating function of the appearance of CCAs among n PCs,

$$Q_n(s) = \sum_{k=1..\infty} CCA_{n,k} s^k.$$

The probability-generating function of the frequency of individuals with CCAs and PCs is described by

$$W(s,t) = \sum_{n=1..\infty} Q_n(s) * PC_n t^n \qquad (3.4)$$

Binomial law of the occurrence of CCAs among PCs: If we put the probability-generating function for the binomial law $Q_n(s) = (qs + (1-q))^n$ into formula (3.4), then

$$W(s,t) = \sum_{n=1..\infty} (qs + 1 - q)^n PC_n t^n = P((qs + 1 - q)t),$$

where q is the probability of the transformation of PC into CCA. If we put $t = 1$ in formula (3.4), then the probability-generating function of the frequency of individuals with CCAs is

$$W(s,1) = P(qs + (1-q)).$$

---

[6]Earlier, we considered not the binomial, but the Poisson law of the appearance of CCAs (Florko and Korogodina 2007; Korogodina and Florko 2007). The Poisson law corresponds to the case of a great number of PCs, among which a rare number of CCAs (k) appeared. The binomial law describes the case, when the number of PCs (n) is low. Poisson (n is great) and Gauss (n, k are great) laws can be obtained by the limiting process from the binomial law (Feller 1957).

Possible types of distribution of individuals on the number of CCAs: Let us find conditions when a probability distribution is the distribution of individuals on the number of CCAs at the given binomial law of the appearance of CCAs. It is the given probability distribution with probability-generating function $W(s) = \sum_{n=0..\infty} W_n s^n$. If it is a distribution of individuals on the number of CCAs, then, as it is shown above, $W(s) = W(s,1) = P(qs + (1-q))$. It follows from the following formula:

$$P(t) = W\left(\frac{1}{q}t + 1 - \frac{1}{q}\right).$$

$P(t)$ is the probability-generating function of the probability distribution, if necessary and sufficient conditions are fulfilled (Feller 1957):

(i) for all n, derivatives, $P^{(n)}(t) = W^{(n)}\left(\frac{1}{q}t + 1 - \frac{1}{q}\right)\frac{1}{q^n} \geq 0$, $0 \leq t \leq 1$;
(ii) $P(1) = W(1) = 1$.

To hold condition (i), it is sufficient to put value $q \approx 1$. At $q = 1$ this condition is fulfilled because it is fulfilled for function W and, due to continuity, it is fulfilled in the interval of $q_0 < q < 1$. Condition (ii) is fulfilled at all values of q. It means that any distribution can be a frequency distribution of individuals on the number of CCAs at the binomial law of the appearance of CCAs with parameter q.

Let us investigate the condition that is necessary to impose on the distribution of individuals on the number of PCs, which would lead to binomiality of the distribution on the number of CCAs and its dependence only on the mean of the distribution on the number of PCs.

Distribution of individuals on the number of CCAs verges towards the distribution of individuals on the number of PCs: When the transformation probability from PC into CCA is $q \approx 1$, then $W(s,1) = P(qs + (1-q)) \approx P(s)$. So, the distribution of individuals on the number of CCAs is the same as on the number of PCs.

Distribution of individuals on the number of CCAs depends on the average number of PCs: If dispersion of a distribution of individuals on the number of PCs is not significant, then generating-functions of the occurrence frequency of individuals on the numbers of PCs and CCAs are $P(t) \approx t^m$ and $W(s,1) \approx (qs + (1-q))^m$, respectively, where m is an average number of PCs. We can conclude that the distribution of individuals on the number of CCAs is binomial and its mean value $m_{AK} = mq$.

Let us study the infinitesimality condition of dispersion. In the spectrum analysis, the Relay condition of lines solvability is known as $R > 1.5d$ (Kosarev 2008), where R is a distance between the lines in the spectrum, and d is a dispersion of the spectrum line. In our case, the term "spectrum" means the distribution of individuals on the number of CCAs, and "line" implies a group of individuals with an equal value of the number of PCs for all individuals. So, the line for individuals with number m_{PC} of PCs corresponds to the binomial distribution of individuals on the

3.4 Statistical View on the Cytogenetic Investigations of Instabilities

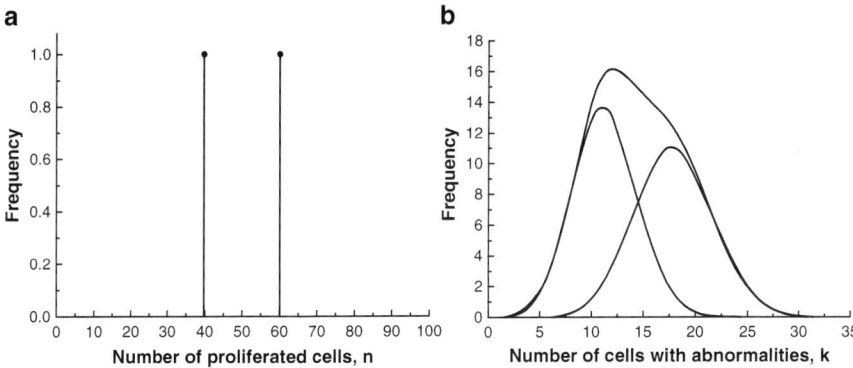

Fig. 3.12 Example of separated groups of individuals with a different number of PCs m_{PC_1} and m_{PC_2} in the distribution of individuals on the number of CCAs. Values $m_{PC_1} = 40$, $m_{PC_2} = 60$, $q = 0.3$. The Relay condition is fulfilled: $\sqrt{m_{PC_2}} - \sqrt{m_{PC_1}} \approx \sqrt{(1-q)/q}$

number of CCAs, and its mean and dispersion values are $m_{CCA} = m_{PC}q$, and $d_{CCA} = \sqrt{m_{PC}q(1-q)}$, respectively. The Relay criterion (see (Kosarev 2008)) was developed for Lorents distribution, which has more sharpness (more excess) than the binomial one. Taking this into account, we offer a modification of the Relay criterion for the binomial distribution: $R > 2d$. The Relay condition looks like $m_{PC_2}q - \sqrt{m_{PC_2}q(1-q)} > m_{PC_1}q - \sqrt{m_{PC_1}q(1-q)}$, which can be rewritten as

$$\sqrt{m_{PC_2}} - \sqrt{m_{PC_1}} > \sqrt{(1-q)/q} \qquad (3.5)$$

Let the distribution of individuals on the number of PCs be characterized by mean m_{PC} and dispersion d_{PC}. If we replace this distribution by two lines (groups of individuals) with the number $R_1 = m_{PC} - d_{PC}$ and $R_2 = m_{PC} + d_{PC}$ of PCs, formula (3.5) becomes:

$$\sqrt{m_{PC} + d_{PC}} - \sqrt{m_{PC} - d_{PC}} < \sqrt{(1-q)/q} \qquad (3.6)$$

We have obtained the Relay condition of invisibility of two groups of individuals for distribution of individuals on the number of PCs and given values of the numbers of PCs (see (Kosarev 2008)). The Relay condition (3.6) determines the value of dispersion d_{PC} of the distribution of individuals on the number of PCs that the distribution of individuals on the number of CCAs would be binomial and depends only on the mean value m_{PC} (Fig. 3.12).

For the binomial law of the appearance of CCAs among PCs, we can assume:

- the distribution of individuals on the number of CCAs will be similar to the distribution of individuals on the number of PCs, if the probability of the appearance of CCAs approaches 1;

- the distribution of individuals on the number of CCAs is binomial and depends only on the mean value of PCs, if the Relay condition is fulfilled (infinitesimality condition of dispersion of the distribution on the number of PCs).

Geometric law of cells with CA appearance. Possible types of the distribution of individuals on the number of CCAs: Let us consider the condition that the probability distribution of individuals would be the geometric distribution of the individuals on the number of CCAs and would be independent of the distribution of individuals on the number of PCs. It is given a probability distribution with generating function $\sum_{n=0..\infty} W_n s^n$. If it is the distribution of individuals on the number of CCAs, then

$$\sum_{n=0..\infty} W_n s^n = W(s,1) = \sum_{n,k=0..\infty} PC_n CCA_{n,k} s^k$$

$$= \sum_{n=0..\infty} PC_n \left(\sum_{k=0..\infty} CCA_{n,k} s^k \right) = \sum_{n=0..\infty} PC_n \frac{1-(qs)^n}{1-qs} \frac{1-q}{1-q^n}.$$

Rewrite it as

$$(1-qs) \sum_{n=0..\infty} W_n s^n = \sum_{n=0..\infty} PC_n \frac{1-(qs)^n}{1-q^n}(1-q),$$

or

$$\sum_{n=0..\infty} (W_n - qW_{n-1}) s^n = \sum_{n=0..\infty} PC_n \frac{1-q}{1-q^n} - \sum_{n=0..\infty} PC_n \frac{(1-q)q^n}{1-q^n} s^n.$$

It means that two conditions would be fulfilled:

$$\begin{cases} qW_{n-1} - W_n = \dfrac{(1-q)q^n}{1-q^n} PC_n \\ W_0 = \sum_{n=0..\infty} PC_n \dfrac{1-q}{1-q^n} \end{cases}$$

From the first condition we have

$$PC_n = (qW_{n-1} - W_n) \frac{1-q^n}{(1-q)q^n}.$$

Hence, realization of the second condition follows identically:

$$\sum_{n=0..\infty} PC_n \frac{1-q}{1-q^n} = \sum_{n=0..\infty} (qW_{n-1} - W_n) \frac{1}{q^n} = W_0.$$

If $PC_n \geq 0$ so $qW_{n-1} - W_n \geq 0$.

3.4 Statistical View on the Cytogenetic Investigations of Instabilities

Hence, $\frac{W_{n-1}}{W_n} \geq q < 1$, or $W_n < W_{n-1}$.

We have found the necessary and sufficient conditions that the given probability distribution W_n could be the distribution of individuals on the number of CCAs with the geometric law of the appearance of CCAs: distribution must monotonically decrease $W_n < W_{n-1}$ and $\sup\left(\frac{W_{n-1}}{W_n}\right) < 1$. To have a distribution of the individuals on the number of PCs, it is sufficient to calculate PC_n by means of the formula $PC_n = (qW_{n-1} - W_n)\frac{1-q^n}{(1-q)q^n}, n = 1..\infty$ for any q satisfying the condition $\sup\left(\frac{W_{n-1}}{W_n}\right) \leq q < 1$.

Let us study how the distribution of individuals on the number of PCs influences the distribution of individuals on the number of CCAs with the geometric law of the appearance of CCAs.

The condition given to the probability distribution of individuals on the number of PCs which leads to the geometric distribution of the individuals on the number of CCAs and its independence on the distribution of individuals on the number of PCs: This condition is held for the number $n \geq 3$ of PCs if the probability of transformation of PC into CCA $q^n \leq 0.125$.[7] Then $CCA_{n,k} = \frac{1-q}{1-q^n}q^k \approx (1-q)q^k$, i.e., the probability of the appearance of CCAs depends on the number k of CCAs only and does not depend on the number n of PCs. Therefore, if the frequency of individuals with the number $n < 4$ of PCs is small, we can neglect the dependence of the appearance law of CCAs on the number of PCs: $CCA_{n,k} = CCA_k$. Then

$$W(s,t) = \sum_{n,k=1..\infty} PC_n CCA_{n,k} t^n s^k = \sum_{n,k=1..\infty} PC_n t^n CCA_k s^k$$

$$= \left(\sum_{n=1..\infty} PC_n t^n\right)\left(\sum_{k=1..\infty} CCA_k s^k\right) = P(t)Q(s).$$

It means that $W(s,1) = P(1)Q(s)$.

We can conclude that the distribution of the individuals on the number of CCAs does not depend on the appearance law of PCs, if $q < 0.5$ and the frequency of individuals with the number $n < 4$ of PCs is small.

Distribution of individuals on the frequency of cells with abnormalities in blood lymphocytes: The distribution of individuals on the frequency of CCAs among the PCs $\rho = \frac{k}{n}$ can be determined when evaluating k as $k = \rho n$ in the distribution of individuals on the number of CCAs:

$$CCA(\rho) = \sum_{n=1..\infty} PC_n * CCA_{n,n\rho}.$$

---

[7] It is $q < 0.5$ in investigations of seeds of plantain populations, and $q < 0.1$ – in studies of blood lymphocytes of individuals (Korogodina et al. 2010a, b).

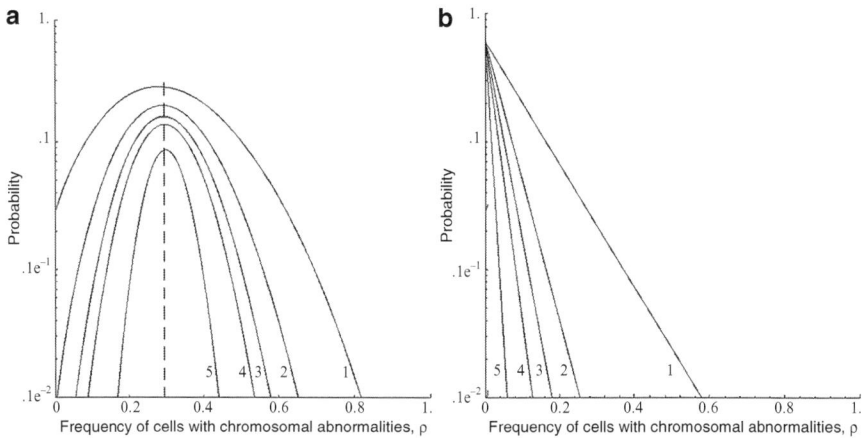

Fig. 3.13 Functions $CCA(\rho)$ for binomial (**a**) and geometric (**b**) laws of the appearance of CCAs among the PCs. Parameter values: $q = 0.3; n = 10$ (pl.1), $n = 20$ (pl.2), $n = 30$ (pl.3), $n = 40$ (pl.4), $n = 100$ (pl.5)

We can find the distribution of individuals on the frequency $CCA(\rho)$ for the binomial law of the appearance of CCAs by the Stirling formula (Fig. 3.13a):

$$CCA(\rho) = CCA_{n,n\rho} = \frac{1}{\sqrt{2\pi\rho(1-\rho)n}} \left[\frac{q}{\rho}\frac{1-q}{1-\rho}\right]^n e^{-\frac{\theta}{12n}}, 0 < \theta < 1 \quad (3.7)$$

The distribution of individuals on the frequency of for the geometric law of the appearance of CCAs can be determined in a similar manner (Fig. 3.13b):

$$CCA(\rho) = CCA_{n,n\rho} = \frac{1-q^n}{1-q} q^{n\rho}. \quad (3.8)$$

Figure 3.13 shows functions for $CCA(\rho)$ different n and $q = 0.3$. Two features of the distribution of individuals on the frequency of CCAs are important.

For different n, all binomial distributions of the frequency of CCAs have a common maximum at $\rho_{max} = q$ (Eq. 3.5, Fig. 3.13a) resulting in sharpening of the distribution of individuals on the frequency of CCAs.

The geometric distribution of individuals on the number of CCAs is independent of the number of PCs, but its conversion into the distribution of individuals on the frequency of CCAs transforms it into a combination of the exponents dependent on the numbers n (Eq. 3.6, Fig. 3.13b) of PCs.

So, conversion to the distribution of individuals on the frequency of CCAs sharpens the distribution of individuals on the number of CCAs in the case of the binomial law (Fig. 3.13a), and enlarges in the case of the geometric law (Fig. 3.13b). It can be suggested that the geometric distribution of individuals on the frequency of CCAs is a character of the geometric law of the appearance of CCAs among the PCs, and the individuals are geometrically distributed on the number of CCAs. The

3.4 Statistical View on the Cytogenetic Investigations of Instabilities

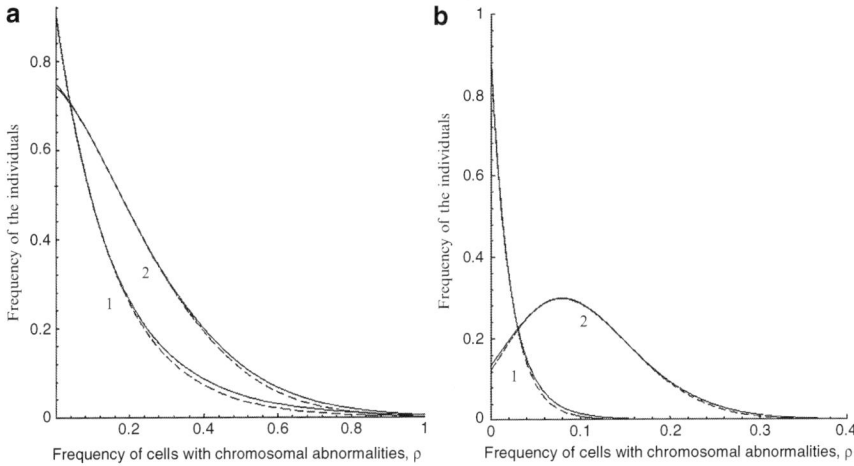

Fig. 3.14 Distribution of the individuals on the frequency of CCAs at the lognormal distribution of individuals on the number of PCs (*the solid line*) and fixed number of PCs, equal to the LN distribution mean (*the dotted line*). Parameter values of the geometric (*1*) and binomial (*2*) laws: $q=0.1, \mu=1, \sigma=0.1$ (**a**) and $q=0.1, \mu=3, \sigma=0.1$ (**b**)

bell-shaped distribution of the individuals on the frequency of CCAs is a character of the binomial law of the appearance of CCAs among the PCs.

Let us consider how the characters of the distribution of the individuals on the number of PCs influence the distribution of the individuals on the frequency of CCAs.

The case when the distribution of individuals on the number of CCAs is independent of the distribution of individuals on the number of PCs: If the frequency of individuals with a small number of PCs is not significant, the distribution of the individuals on the frequency of CCAs does not differ from the distribution of individuals on the frequency of CCAs for the group of individuals with a fixed number of PCs (Fig. 3.14a, b). Distributions on the frequency of CCAs and on the number of CCAs are appropriate equally to analyze the binomial law as well as the geometric one of the appearance of CCAs.

The case when the distribution of the individuals on the number of CCAs depends on the distribution of the individuals on the number of PCs: At the conversion to frequency, in the case of the binomial law of the appearance of CCAs among the PCs, the distribution of the individuals on the number of CCAs converges to the distribution of individuals on the frequency. In the case of the exponential law of the appearance of CCAs among the PCs, the distribution of the individuals on the frequency of CCAs obtains a "tail" and is broader (Fig. 3.15a, b). We can conclude that the distribution of the individuals allows one to determine accurately parameter q of the binomial law of the appearance of CCAs among the PCs, but geometric parameter q can be correctly calculated by using the distribution of individuals on the number of CCAs.

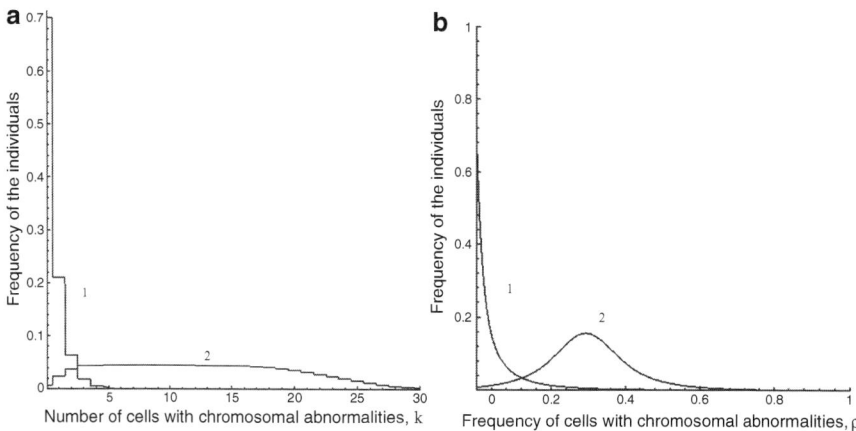

Fig. 3.15 Distribution of the individuals on the number of CCAs (**a**), distribution of the individuals on the frequency of CCAs (**b**). *1* – the geometric law, *2* – the binomial law

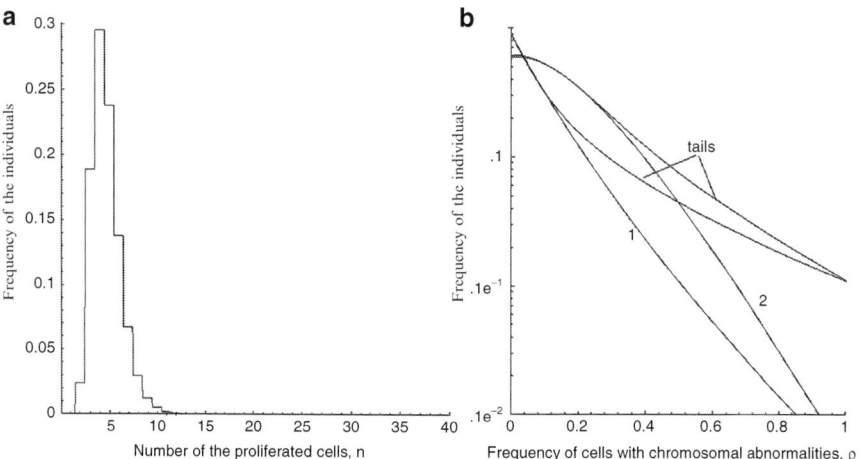

Fig. 3.16 The lognormal distribution of individuals on the number of PCs, mean $m_{PC} = 5$ (**a**). Completed distribution of the individuals on the frequency of CCAs for the geometric (*1*) and binomial (*2*) laws of the appearance of CCAs among the PCs (**b**). Parameter $q = 0.3$

Let us consider how violation of the infinitesimality condition of the occurrence of the individuals with small numbers of PCs influences the frequency distribution of the individuals. The distribution of the individuals on the frequency of CCAs gets a long tail in the region of the great CCA frequencies, which is related to the above broadening of the geometric distribution of the individuals on the number of CCAs (Fig. 3.16b).

Comparison of preliminary conclusion with the experimental distributions: Let us examine our conclusions by comparison of the statistical modelling with

3.5 Some Conclusions

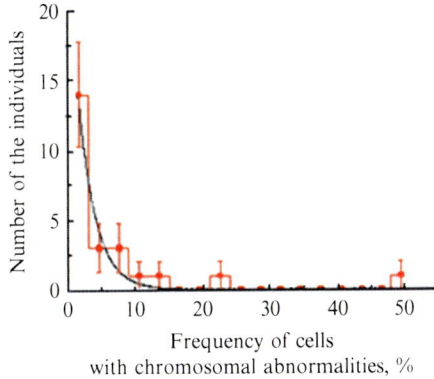

Fig. 3.17 The experimental distribution of individuals on the frequency of CCAs for the individuals whose number of activated cells of lymphocytes didn't exceed 30 ($N_{mph} < 30$). Samples of the individuals living in the Tyumen and Irkutsk regions

the experimental data on blood lymphocyte samples of the individuals living in settlements in Samburg (Tyumen region), Maloe Goloustnoe (Irkutsk region), and the city of Novosibirsk (Korogodina et al. 2010a). Distributions of the individuals on both the number of PCs and the CCA frequency are shown in Figs. 3.17 and 3.18.

Figure 3.17 presents the experimental distribution of the individuals on the frequency of CCAs for individuals whose sample of blood contained a small number of proliferated lymphocyte cells ($N < 30$) (Antonova et al. 2008). We assume that the distribution for this group of individuals has a long tail because the requirement of small numbers of individuals with small numbers of PCs is not fulfilled.

In the cases when the Relay conditions are preserved for the distributions of the individuals on the number of PCs (Fig. 3.18b–d), the Poisson distribution of the individuals on the frequency of CCAs is observed. Contrarily, the individuals are not Poisson-distributed on the frequency of CCAs (Fig. 3.18a) in the case when the Relay conditions are not preserved. According to our conclusions, the type of distribution of individuals on the frequency of CCAs has to be the same as the type of distribution of individuals on the number of CCAs.

3.5 Some Conclusions

1. Distributions of individuals on the number of CCAs and the frequency of CCAs follow the binomial law under the conditions that (1) the law of the appearance of CCAs among the PCs is binomial and (2) the Relay condition is fulfilled for the distributions of the individuals on the number of PCs. Then, the distribution of the individuals on the number of CCAs is independent of the characters of the distribution of the individuals on the number of PCs.
2. Distributions of the individuals on the number of CCAs and the frequency of CCAs are geometric and independent of the type of distribution of individuals on the number of PCs under the condition that the law of the appearance of CCAs among the PCs is geometric, the parameter of geometrical distribution $q < 0.5$ and the probability of the cases of individuals with number of PCs $n < 4$ is small.

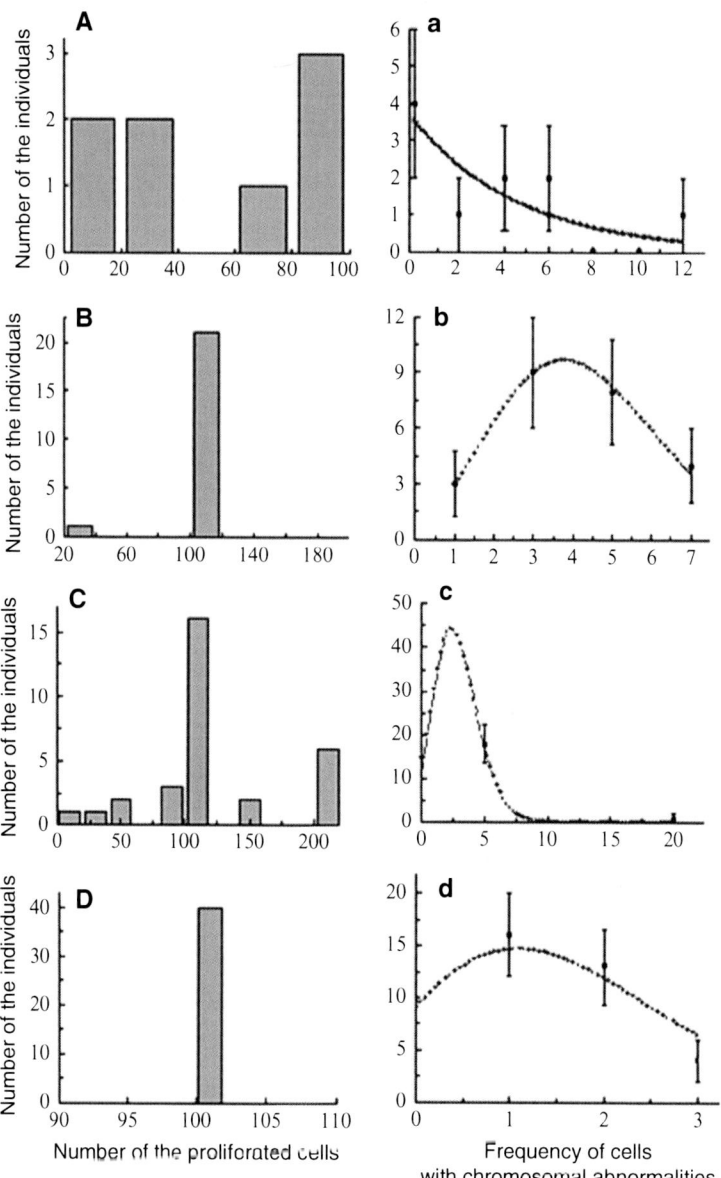

Fig. 3.18 The distribution of the samples of the individuals on the number of PCs (**A–D**) and on the frequency of CCAs (**a–d**). Settl. Maloe Goloustnoe: 60–80 (**Aa**) and <40 (**Bb**) years old; Samburg: <18 years old (**Cc**); (**c**) Novosibirsk, control (**Dd**)

3. The exponential distribution of the individuals on the frequency of CCAs is an indicator of the geometric law of the appearance of CCAs among the PCs. It is also an indicator of the geometric distribution of the individuals on the number of CCAs. The bell-shaped distribution of individuals on the frequency of CCAs is the indicator of the binomial law of the appearance of CCAs among the PCs.
4. Distributions of the individuals on the number of CCAs and the frequency of CCAs can be used to determine the type and parameters of the law of the appearance of CCAs among the PCs. To calculate the parameter q, the analysis of the geometric distribution of the individuals on the number of CCAs can be effectively used, whereas the binomial distribution on the frequency of CCAs should be investigated.
5. The geometric distribution of individuals on the frequency of CCAs among PCs can be an indicator of termination of the process of the appearance of CCAs whereas the bell-shaped distribution indicates its absence. Within the scope of the adaptation model, we can expect the selection processes in the case of geometric distribution to be in agreement with the investigations of H.A. Orr who has shown that the distribution tail is related to selection in the evolution model (Orr 2006).

3.6 Summary

The statistical modelling can be used in the cases when experimental material is based on the events the occurrence of which follows different laws. The radiation-induced adaptation processes can be studied by this method.

Adaptation process, its representation for seeds population: Adaptation process has three components: the primary and late cell damage processes, and selection. They are different in the resistant and sensitive subpopulations: the transformation probability of the normal cell into the abnormal one, as well as its death, is higher in the sensitive fraction. The selection process characterizes the sensitive seedlings. In the resistant fraction, the abnormalities are accumulated without selection (Korogodina et al. 2010a).

The geometric law describes the processes terminated by the "success" (Feller 1957). In our case, it means successful selection: the series of late damages break at the seedling adaptation. The Poisson law describes the rare and independent appearance of events, which are the primary and secondary damages of meristem cells in the resistant seedlings. So the adaptation process can be described by the sum of the Poisson and geometric (P + G) laws (Florko and Korogodina 2007).

Representation of adaptation process in cells and its regularities: We can expect the same scheme of adaptation in cell populations: the chromosomal abnormalities appear in cells as independent and correlated events, and the repair system takes part in the cells' selection. The appearance of the independent injuries relates to the Poisson law. The selection process means correlation of the damages occurrence, and can be described by the geometric law. The type and parameters of distributions depend on the intensities of the primary and late damages, as well as the repair rate (Florko and Korogodina 2007).

The constant adaptation process occurs in the intact cells that reveals as the geometric distribution (G) on the chromosomal abnormalities (Korogodina et al. 2010a). The radiation stress factor induces a more intensive adaptation process in the sensitive cell subpopulation. It is described by the second geometric distribution related to a more sensitive group of cells (G1 + G2).

The primary damages frequency increases in the cell while the irradiation intensity increases. If the rate of the primary damage appearance exceeds the generation rate of the geometric distribution tails, then the primary damages appearing earlier should generate the Poisson distribution according to the queues theory[8] (G + P). In addition to that, the quota of the geometric distribution will decrease. It is important that the type of distribution depends only on the primary damage intensity.

These conclusions allow us to use the geometric and Poisson distributions to describe the appearance of chromosomal abnormalities in cells.

Representation of the proliferated cells occurrence: A.N. Kolmogorov studied the lognormal distribution of particle sizes under fragmentation. It has been shown that distribution of the particle number is described by the lognormal law (Kolmogorov 1986). The cell reproduction process can be presented as the fragmentation, and then the cell multiplication can be described by the lognormal law.

Radiation stress induces cell death and stimulates the resting cells to proliferation. How do these processes change the seeds distributions on the number of proliferated cells? The seed population is heterogenic and consists of resistant and sensitive subpopulations. It means that the experimental distribution is divided into two groups with different parameters reflecting the seed properties (LN1 + LN2). The stimulation of cell proliferation adds the third lognormal component in the experimental distribution of proliferated cells (LN1 + LN2 + LN3). The strong radiation stress can transform the lognormal law into the geometric one (LN1 + LN2 + G).

Formation of the individual distribution on the occurrence frequency of abnormal lymphocyte cells among the proliferated ones: The occurrence frequency of lymphocyte cells with abnormalities among the proliferated ones is the ratio of the number of abnormal cells to the normal ones. So the process of abnormal cells appearance determines the type of individuals' distribution on the occurrence frequency of abnormal lymphocyte cells among the proliferated ones. It can be Poisson/binomial or geometric laws (Florko et al. 2009).

Therefore, the exponential distribution of the individuals on the frequency of cells with abnormalities is an indicator of the geometric law of the abnormal cells appearance among the proliferated ones. It is also an indicator of the geometric

---

[8]See the general queues theory in (Feller 1957). A.N. Chebotarev used the queues theory to study the appearance of multiaberrant cells (2000).

distribution of the individuals on the abnormal cells number. The bell-shaped distribution of individuals on the abnormal cells frequency is the indicator of the binomial law of the abnormal cells appearance among the proliferated ones.

The geometric distribution of individuals on the frequency of cells with abnormalities among the proliferated ones can be an indicator of termination of the process of the abnormal cells appearance (it means selection), whereas the bell-shaped distribution indicates its absence. Distributions of the individuals on the abnormal cells numbers and the abnormal cells frequency can be used to determine the type and parameters of the law of the abnormal cells appearance among the proliferated ones.

References

Antonova E, Osipova LP, Florko BV et al (2008) The comparison of distributions of individuals with normally and poorly stimulated blood cell activity on the frequency of aberrant cells' occurrence in blood lymphocytes. Rep Russ Mil-Med Acad 1(3):73 (Russian)

Arutyunyan R, Neubauer S, Martus P et al (2001) Intercellular distributions of aberrations detected by means of chromosomal painting in cells of patients with cancer prone chromosome instability syndromes. Exp Oncol 23:23–28 (Russian)

Bochkov NP, Chebotarev NA (1989) Human heredity of mutagens of environment. Medicina, Moscow

Bochkov NP, Yakovenko KN, Chebotarev AN et al (1972) Distribution of the damaged chromosomes on human cells under chemical mutagens effects in vitro and in vivo. Genetika 8:160–167 (Russian)

Bridges BA (1997) DNA turnover and mutation in resting cells. Bioessays 19(4):347–352

Chebotarev AN (2000) A mathematical model of origin of multi-aberrant cell during spontaneous mutagenesis. Rep RAS 371:207–209 (Russian)

Condit R, Hubbell SP, La Frankie JV et al (1996) Species–area and species–individual relationships for tropical trees: a comparison of three 50-ha plots. J Ecol 84:549–562

Feller W (1957) An introduction to probability theory and its applications. Wiley/Chapman & Hall, Limited, London/New York

Fisher RA (1930) The genetical theory of natural selection. Oxford University Press, Oxford

Fisher RA, Corbet AS, Williams CB (1943) The relation between the number of species and the number of individuals in a random sample of an animal population. J Anim Ecol 12:42–58

Florko BV, Korogodina VL (2007) Analysis of the distribution structure as exemplified by one cytogenetic problem. PEPAN Lett 4:331–338

Florko BV, Osipova LP, Korogodina VL (2009) On some features of forming and analysis of distributions of individuals on the number and frequency of aberrant cells among blood lymphocytes. Math Biol Bioinform 4:52–65, Russian

Gillespie JH (1983) A simple stochastic gene substitution model. Theor Popul Biol 23:202–215

Gillespie JH (1984) Molecular evolution over the mutational landscape. Evolution 38:1116–1129

Gnedenko BV (1965) The course of probability theory. Nauka, Moscow (Russian)

Harris TE (2002) The theory of branching processes. Courier Dover Publications, New York

van Kampen NG (2007) Stochastic processes in physics and chemistry. Elsevier, Amsterdam

Kimura M (1983) The neutral theory of molecular evolution. Cambridge University Press, Cambridge

Klauder J, Sudarshan E (1968) Fundamentals of quantum optics. Benjamin, New York

Kolmogorov AN (1986) About the log-normal distribution of particle sizes under fragmentation. In: The probabilities theory and mathematical statistics. Nauka, Moscow (Russian)

Korogodina VL, Florko BV (2007) Evolution processes in populations of plantain, growing around the radiation sources: changes in plant genotypes resulting from bystander effects and chromosomal instability. In: Mothersill C, Seymour C, Mosse IB (eds) A challenge for the future. Springer, Dordrecht, pp 155–170

Korogodina VL, Florko BV, Osipova LP (2010a) Adaptation and radiation-induced chromosomal instability studied by statistical modeling. Open Evol J 4:12–22

Korogodina VL, Florko BV, Osipova LP et al (2010b) The adaptation processes and risks of chromosomal instability in populations. Biosphere 2:178–185 (Russian)

Korogodina VL, Panteleeva A, Ganicheva I et al (1998) Influence of dose rate gamma-irradiation on mitosis and adaptive response of pea seedlings' cells. Radiat Biol Radioecol 38:643–649 (Russian)

Kosarev EL (2008) Methods of the experimental data processing. Physmatlit, Moscow (Russian)

Lea DE (1946) Action of radiations on living cells. Cambridge University Press, Cambridge

Lorimore SA, Wright EG (2003) Radiation-induced genomic instability and bystander effects: related inflammatory-type responses to radiation-induced stress and injury? A review. Int J Radiat Biol 79:15–25

Luchnik NV (1958) Influence of low-dose irradiation on mitosis of pea. Bull MOIP Ural Department 1:37–49 (Russian)

Luchnik NV (1968) Biophysics of the cytogenetic damages and genetic code. Medicina, Leningrad

McGill BJ, Etienne RS, Gray JS et al (2007) Species abundance distributions: moving beyond single prediction theories to integration within an ecological framework. Ecol Lett 10:995–1015

Morgan WF (2003) Non-targeted and delayed effects of exposure to ionizing radiation. II. Radiation-induced genomic instability and bystander effects in vivo, clastogenic factors and transgenerational effects. Radiat Res 159:581–596

Motomura I (1932) A statistical treatment of associations. Jpn J Zool 44:379–383 (Japanese)

Orr HA (1998) The population genetics of adaptation: the distribution of factors fixed during adaptive evolution. Evolution 52:935–949

Orr HA (1999) The evolutionary genetics of adaptation: a simulation study. Genet Res 74:207–214

Orr HA (2005a) The genetic theory of adaptation: a brief history. Nat Rev Genet 6:119–127

Orr HA (2005b) Theories of adaptation: what they do and don't say. Genetica 123(1–2):3–13

Orr HA (2006) The distribution of fitness effects among beneficial mutations in Fisher's geometric model of adaptation. J Theor Biol 238:279–285

Preston FW (1948) The commonness and rarity of species. Ecology 29:254–283

Preston FW (1962) The canonical distribution of commonness and rarity. Part I. Ecology 43:185–215

Reed WJ, Hughes BD (2002) From gene families and genera to incomes and internet file sizes: Why power laws are so common in nature. Phys Rev E 66:67–103, Article number 067103

Robbins CS, Bystrak D, Geissler PH (1986) The breeding bird survey: its first fifteen years, 1965–1979. US Department of the Interior Fish and Wildlife Service, Washington, DC

Vasiliev AG, Boev VM, Gileva EA et al (1997) Ecogenetic analysis of late consequences of the Totskij nuclear explosion in Orenburg region in 1954 (facts, models, hypotheses). Ekaterinburg, Ekaterinburg

Whittaker RH (1960) Vegetation of the Siskiyou mountains, Oregon and California. Ecol Monogr 30:279–338

Whittaker RH (1965) Dominance and diversity in land plant communities. Science 147:250–260

Winemiller KO (1990) Spatial and temporal variation in tropical fish trophic networks. Ecol Monogr 60:331–367

Chapter 4
Non-linearity Induced by Low-Dose Rates Irradiation. Lab Experiments on Pea Seeds

Abstract General regularities of low dose-rate radiation effects were determined on plant seeds in laboratory experiments. Three groups of seeds were tested: young, old, and heat-stressed because aging and high temperature are the usual natural factors. Statistical modelling has shown that the adaptation processes have three components: primary injuring, late damaging, and selection, which is more intensive in the sensitive subpopulation of seeds. The relationship between inter- and intracellular processes was analyzed. The combination of radiation and heat stresses dramatically increases late damaging and selection at non-optimal temperatures which do not exceed the norm limits. This approach allows one to estimate the risks of instabilities accompanied by the accumulation of abnormalities, selection, and death of seedlings.

Keywords Plant seeds • low dose- and dose-rate irradiation • aging • high temperature • synergic effect • hormetic effect • adaptive response • nonlinear response • statistical modelling • instability • selection • sensitivity • chromosomal abnormalities • inter- and intracellular adaptation processes • risk of instability

4.1 The Seeds of a Plant Are an Important and Convenient Object in Radiobiology and Radioecology

Researches of low-dose radiation effects require the irradiation of a large population in the laboratory. This is impossible in the case of animals or humans. To have sufficient statistics, the seeds of plants can be used to study general regularities of radiation effects in the laboratory.

This is possible because the meristematic cells of plants are analogous in function to stem cells in animals, and the radiation-induced processes in meristem can be a model of the effects in bone marrow. Plants are well-investigated objects; as opposed to animals, and especially to humans, plants and their seeds are more accessible. This object is favorable in radioecological investigations.

4.1.1 The First Investigations of Non-linearity Were Performed on Plants

Nikolay V. Luchnik was the first to show non-linearity of the frequency of cells with abnormalities at low irradiation doses and their stimulation in dividing the resting cells in seedling meristem (Luchnik 1958). This investigation was performed at the laboratory of Nikolay W. Timofeeff-Ressovsky in Miassovo (the South Urals), where the low-dose radiation effects were studied.[1] Luchnik chose the pea seeds because the pea is a genetically well-investigated object and radiosensitive species. In Luchnik's investigations, pea seeds were soaked in β-radiator solution with 0.25–10.00 mCi/l during some hours; then percentages of normal and abnormal mitosis were registered. The results of his experiment (enacted repeatedly with different concentrations of the radiator and different durations of soaking for the seeds) showed that a low concentration 0.5 mCi/l of the radiator induced the greatest stimulation of mitotic activity and a decrease of the percentage of abnormal mitoses (Table 4.1). Then, the mitotic activity normalized, and the percentage of abnormal mitoses gradually increased. So, the non-linearity of both mitotic activity and the frequency of chromosomal abnormalities were demonstrated at the low-dose and dose-rate irradiation.

The experiments of Luchnik were repeated at the Joint Institute for Nuclear Research after the Chernobyl accident when interest in the effects of low-dose radiation increased (Korogodina et al. 1998). The concentrations of β-radiator used by Luchnik were recalculated into dose-rate γ- irradiation. The results of the experiments were in agreement with the conclusions made by Luchnik. Figure 4.1 shows a standard dose-dependence of the frequency of cells with abnormalities in rootlet meristem and survival of irradiated seeds (Fig. 4.1a), and the same characteristics after the pea seeds' irradiation at the dose and dose rates used by Luchnik (Fig. 4.1b) (Korogodina et al. 1998).

Low-dose irradiation of seeds, seedlings or vegetative plants can usually induce the so-called effect of radiation stimulation, which appears as the acceleration of seeds sprouting, the growth of root and sprout, and propagation of the stage

---

[1] After the Second World War, Timofeeff-Ressovsky was sent by the regime first to a Stalinist camp near Karaganda, and then to Site No. 0215, a secret scientific laboratory in the South Urals. There, he headed pioneering studies of the biological effect of radionuclides and their impact on ecosystems. Many investigators who later gained scientific repute worked at the Timofeeff-Ressovsky laboratory; the others came there for discussion. Among these scientists were V.I. Korogodin, G.G. Polikarpov, A.A. Lyapunov and others.

Table 4.1 Influence of different concentrations of β-radiator solution on mitotic activity and percentage of abnormal mitoses during 70 h after pea seeds' soaking (Luchnik 1958)

Concentration, mCi/l	Dose, rep[a]	Mitoses, %	Abnormal mitoses, %
Control	0	10.6 ± 0.25	4.5 ± 0.7
0.25	7	29.3 ± 0.59	13.3 ± 2.2
0.5	14	43.5 ± 0.42	10.2 ± 1.1
1.0	27	24.4 ± 0.38	14.7 ± 1.3
2.0	54	19.8 ± 0.48	20.7 ± 1.8
5.0	125	11.6 ± 0.83	14.7 ± 2.9
10.0	250	11.6 ± 0.83	22.0 ± 3.4

[a]Roentgen-equivalent-physical

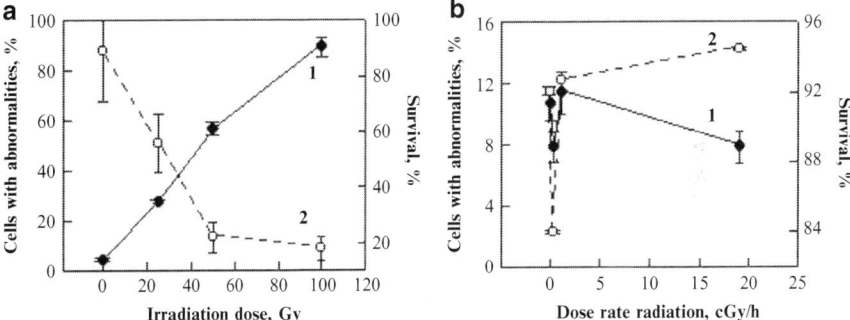

Fig. 4.1 Investigations of dose-dependence of the pea (*Pisum arvense*) seed survival and frequency of cells with abnormalities in rootlet meristem of seedlings of irradiated seeds (**a**) and dependence of the same characteristics at the dose and dose-rate γ-irradiation of pea seeds corresponding to the Luchnik experiments (**b**). *1*, frequency of meristem cells with abnormalities; *2*, survival of seeds

development. Stimulation doses vary widely for different plant species. Acceleration of growth of the plant is a consequence of the increasing of cells' division: for example, γ-irradiation at 0.35 Gy of pea seedlings increases the relative growth rate and mitotic activity of the seedlings' meristem cells as well as reducing the duration of the cell cycle (Gudkov 1985).

4.1.2 Modification of Radioresistance of Plants

Temperature: Temperature is a standard physical factor, the increasing and decreasing of which influences seeds in different ways. It is a well-known fact that high temperature increases seeds' germination, while low temperature increases the survival of seeds in storage. The latter can be explained by a low level of the free radicals which can accumulate under high temperatures.

It is important to consider the synergic radiation and high temperature effects because it is a standard condition in nature. V.G. Petin and his colleagues investigated synergism in detail (Petin et al. 1999, 2002; Kim et al. 2001). The authors obtained a bell-shaped curve for synergic interaction of heat and UV light, UV light and methotrexate, UV light and cadmium chloride, X-rays and chemical mutagens, γ-rays and hyperthermia, UV light and hyperthermia, and finally ultrasound and hyperthermia. It is most probable that the bell-shaped curve reflects a general feature of synergic interaction. The authors offered a hypothesis that synergism is a result of some additional lethal lesions arising from interaction of the sublesions induced by both factors separately because synergic interaction is not related to direct damage. Singly, these sublesions are not necessarily lethal. The investigations of other authors have verified this hypothesis (Dineva et al. 1993).

Humidity: Humidity plays an important part in seeds' radioresistance as demonstrated by many scientists (reviewed in (Gudkov 1985)). Experimental data on many cultures indicated that the highest radioresistance of seeds is observed at water concentration much higher than in air-dry seeds (Ohba 1961). Researchers have come to the conclusion that the water content is an important factor which regulates the level of free radicals not only in the irradiation moment but also in a post-radiation period. One can assume that the role of humidity is not only related to the free radical mechanism. It is known that the water content in a cell influences the metabolic processes, conformation of the macromolecules, and etc.

The growth factors of plants exert a different influence on the radiation injuring and post-radiation recovery. Their efficiency depends on the moment of action: before irradiation, in the moment of radiation, or in a post-radiation period. Investigations of the influence of chemical factors-modifiers of seed radioresistance are reviewed in (Gudkov 1985; Atayan 1987; Evseeva and Geras'kin 2001), and they will not be considered here.

4.1.3 Meristems Are Critical Tissues of Plants

Plants have a unique feature: the ability for non-limited growth due to the embryonic tissues, in their meristem. The main feature of meristem is the existence of cells which preserve an ability to produce both divided and differentiated cells.

Apical meristems contain some thousands of cells at the caps of sprouts and roots. Root meristem is characterized by a relatively simple structure, discrete separations of elongation and differentiation, and high proliferative activity zones. This object is very suitable for the study of growth processes and cell reproduction.

Figure 4.2 presents schemes of the root apex. It shows some zones, including the meristem zone (zone of division) and zone of elongation. The depot of all types of cells is a zone of initial cells which consists of juvenile constantly proliferated

4.1 The Seeds of a Plant Are an Important and Convenient Object in Radiobiology...

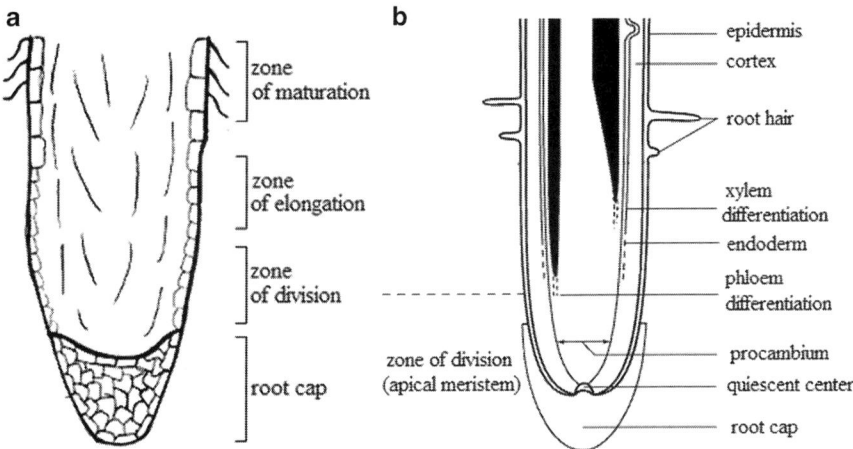

Fig. 4.2 Structure of the apical zone of seedling's root

cells. The initial cells are typical stem cells characterized by an ability to continue indefinitely, produce differentiated cells and recover a normal number of cells after damaging. Initial cells form a population which then passes into differentiation and specialization. The zone of initial cells is located at the board of a quiescent center and numbers approximately a hundred cells (a thousand for a pea). Cells of the quiescent center could be considered stem cells, too, although these cells divide by a rate 10–30 times slower. Initial cells divide down (distal) and form a root cap which protects meristem cells against damages.

So, the indefinite ability of initial cells to divide, the availability of the proliferative subpopulation, and the ability of their cells to differentiate and then specialize equate meristem cells to the renewed tissues of the animal such as blood, epithelium of bowels, liver and so on. These tissues were called "critical" due to their high radiosensitivity. In the animal, damage of these tissues leads to marrowy and gastrointestinal syndromes, sickness caused by radiation and death of the organism. The plant meristem is highly sensitive to any damages. Meristem sensitivity is more than 10–100 times higher than for differentiated and specialized tissues. Therefore, radiobiology of high plants is the radiobiology of their meristem.

Meristem is heterogenic relative to its cell sizes, which are the cause of a different duration of the cell mitotic cycle, while a different cycle stage results in different resistance for cells.

Irradiation of cells with middle-lethal doses[2] increases their mitotic cycle duration. The duration of the cell cycle is caused by mitosis delay related to DNA recovery by a radiation-damaged cell (Boei et al. 1996). Perhaps the rapidly divided

---

[2]The interval of middle-lethal doses corresponds to Fig. 4.1a

cells of irradiated meristem are those the cycle phases of which were radioresistant and therefore their cycle duration remained approximately normal.

The repopation processes can be provided due to the existence of heterogenic subpopulations of cells which differ in their proliferative activity and radioresistance. Repopulation recovery is going on due to the reproduction of cells in the radioresistance stages or by resting cells. Such cells can proliferate, and their number determines ability and rate of repopulation recovery (Gudkov 1985). Resting cells are the most resistant to irradiation. It is a non-specific protection of plant organisms.

4.2 Laboratory Experiments on Pea Seeds

The pea seeds were irradiated in a laboratory to study general regularities of the low-dose and dose-rate radiation effects. These lab experiments on seeds were a revision of the experiments performed at the Urals laboratory headed by N.W. Timofeeff-Ressovsky. The same dose and dose-rates irradiation used by N.V. Luchnik (1958) in his experiments were chosen. Radiation effects were studied under modification factors of aging and high temperature. These factors had to imitate natural conditions: the seeds were stored for 20 months, and high temperatures did not exceed ecological norms for the given seeds. These three factors are connected with reactive oxidative species (ROS) production (Buzzard et al. 1998; Pinzino et al. 1999; Seymour and Mothersill 2000; Mothersill et al. 2000b), which is the reason to suspect instability processes.

Statistical modelling was performed to investigate the process of instability and its dependence on dose-rate irradiation, aging, and high temperature.

This series of lab experiments was carried out at the Joint Institute for Nuclear Research (Dubna) using the methods developed at the Vavilov Institute of General Genetics of the Russian Academy of Sciences (Moscow) and the Russian Institute of Agricultural Radiology and Agroecology of the Russian Academy of Agricultural Sciences (Obninsk).

4.2.1 On the Object and Methods Used in Lab Investigations

Pea seeds (*Pisum arvense*), selected line Nemchinovsky-817 (received from Moscow, Nemchinovka, Agriculture Institute), were used in the laboratory experiment. For this kind of seed, the reported quasi-threshold radiation dose which corresponds to inflection of the survival curve from a shoulder to mid-lethal doses is 10–20 Gy (Preobrazhenskaya 1971). The seeds can be divided into three groups on the character of their storage (Korogodina et al. 2005). The first group of seeds ("young") was stored in the refrigerator for 8 months and was tested in April; the second group was kept in the refrigerator for 20 months ("old") and then tested in April; and the third group was stored in the refrigerator for 8 months and then held outdoors until June when it was tested. The third group of seeds was stored without sunlight and precipitation, but the temperature during the day and night reached 30–32°C and 16–18°C, respectively (they are higher for the Moscow region). The duration of heat stress was 2 months ("young seeds with heat stress").

Pea seeds were gamma-irradiated at room temperature with 7 cGy ^{60}Co at 0.3, 1.2, or 19.1 cGy/h. The overall uncertainty of dose rates and doses was 10–13 % at a dose rate of 19.1 cGy/h and ~3–5 % at dose rates of 0.3–1.2 cGy/h. Ana–telophases were scored for chromosomal abnormalities (CAs) containing chromosome bridges and acentric fragments. The mitotic index (MI) of seedling meristem cells was scored as the percentage of cells in mitosis. The mathematical processing is described in Applications 8.1, 8.2.

4.2.2 Occurrence of Cells with Abnormalities at Different Dose-Rate Irradiation. Effects of Aging and High Temperature

Seedlings were fixed when they reached the length corresponding to the first mitosis or were clearly unable to survive. The cells with chromosomal abnormalities (CCAs) and the MI were registered in the sprouts' germination. The times of seeds' germination and occurrence of abnormal and normal cells were different for the tested dose-rates irradiation and three groups of seeds (Fig. 4.3). This is expected because cell cycles can be perturbed by irradiation (Gudkov 1985; Burdon et al. 1989; Boei et al. 1996) that includes the cell cycle delay (Boei et al. 1996) as well as stimulation to proliferation of resting cells (Luchnik 1958; Gudkov 1985). Let us consider the averaged characteristics of seed survival, frequency of cells with abnormalities and the MI (Table 4.2).

Testing of young seeds: In either case, the frequency of CCAs increased with time to fixing in young seeds both without additional irradiation ($p < 0.001$) or irradiated at 19.1 cGy/h ($p < 0.001$) (Fig. 4.3a, d). In the latter case, the first fixation started later in comparison with the control because DNA damage can delay cell cycles (Boei et al. 1996) and the daily seedling growth rate (Gudkov 1985). Cell cycle delays lead to the appearance of delayed cells at the last fixation, elevating their MI values (Fig. 4.4d). It seems that some cells with DNA damage are eliminated (the frequency of CCAs decreased, but not significantly) (Table 4.2). The current observations have indicated that the high-dose rate (19.1 cGy/h) sufficiently delays germination, although the average values of $(1 - S)$, the MI and the frequency of CCAs do not significantly change in comparison with the control.

In contrast, the frequency of CCAs increased more slowly with time to fixing at 0.3 and 1.2 cGy/h than at 0 or 19.1 cGy/h ($p < 0.001$) (Fig. 4.3b, c). It is known that oxidative stress induced by low-dose irradiation (Mothersill and Seymour 2000) causes cell elimination (MacCarrone et al. 2000). These data allow one to surmise

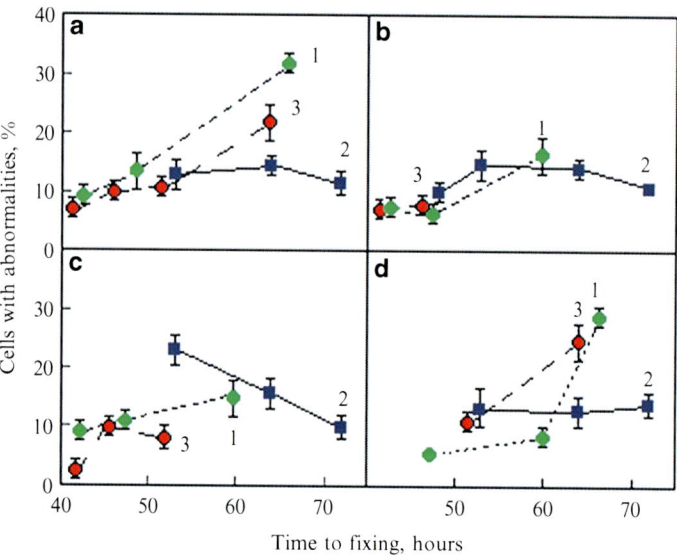

Fig. 4.3 Frequencies of pea seedling meristem CCAs at the first mitosis without irradiation (**a**) and with 7 cGy at a dose rate of 0.3 cGy/h (**b**), 1.2 cGy/h (**c**), and 19.1 cGy/h (**d**). *1*, young seeds; *2*, old seeds; *3*, young seeds with heat stress (heat stress duration 2 months). The time to fixing is the onset of mitosis

Table 4.2 Effects of ionizing irradiation on pea seed populations at different dose rates and storage conditions

Dose rate, cGy/h	N	1 − S, %	CCA %	MI
Young seeds				
0	149	8.0	10.7 ± 1.1	5.8 ± 0.4
0.3	150	16.0*	7.8 ± 1.0*	7.6 ± 0.5*
1.2	150	7.3	11.4 ± 1.4	6.5 ± 0.6
19.1	146	5.5	9.1 ± 1.2	6.0 ± 0.5
Old seeds				
0	125	21.6	13.1 ± 1.0	5.1 ± 0.5
0.3	128	28.1*	12.5 ± 0.9	8.6 ± 0.6*
1.2	174	20.7	15.8 ± 1.4*	5.4 ± 0.4
19.1	149	28.2*	13.1 ± 1.7	4.6 ± 0.5
Young seeds with heat stress				
0	40	7.5	11.2 ± 1.1	5.9 ± 0.3
0.3	40	32.5**	8.2 ± 0.7*	5.5 ± 0.5***
1.2	42	62.5**	6.1 ± 1.0*	6.0 ± 1.2***
19.1	40	50.0**	15.4 ± 2.2*	4.5 ± 0.3*

The standard error is shown. The difference with non-irradiated control: *$p < 0.05$; **$p < 0.001$; ***$p > 0.05$

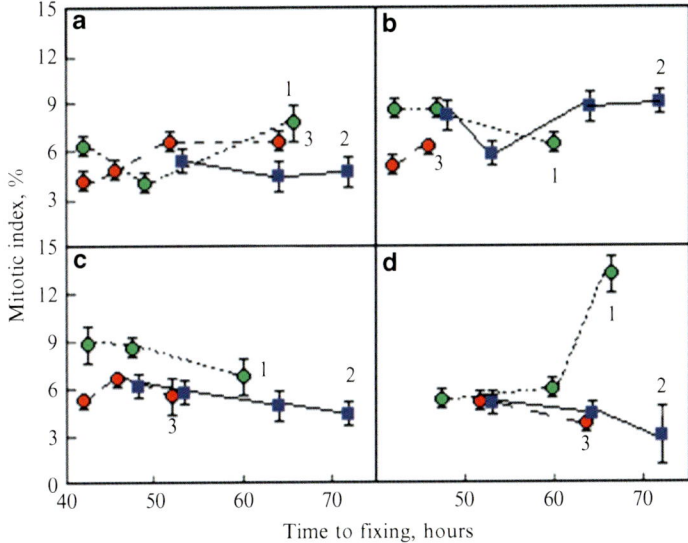

Fig. 4.4 Mitotic index of pea seedling meristem cells at the first mitosis without irradiation (**a**) and with 7 cGy at a dose rate of 0.3 cGy/h (**b**), 1.2 cGy/h (**c**), and 19.1 cGy/h (**d**). *1*, young seeds; *2*, old seeds; *3*, young seeds with heat stress (heat stress duration 2 months). The time to fixing is the onset of mitosis

that irradiation of 0.3 and 1.2 cGy/h increases cell elimination. The result suggests that the slope of both the frequency of CCAs and the MI with time to fixing would decrease, and the fixing period can be reduced. This is the case: no elevation of the MI is observed with time to fixing ($p < 0.05$), and the fixation period is shortening at these dose rates (Fig. 4.3b, c). The hypothesis that the frequency of CCAs decreases due to the error-free repair mechanism cannot be accepted because the $(1 - S)$ value increases (twofold at 0.3 cGy/h) (Table 4.2). An alternative explanation is that the low-radiation dose rates turn on repair mechanisms which lead to cell lethality because such irradiation induces very few damages and unnecessarily expends the cell's resources. However, it is difficult to believe that evolution did not select for organisms' preservation in all cases and did not improve their survival by additional regulation mechanisms at low doses.

In addition, it is necessary to point out the tendency of the MI value to increase at 0.3 ($p < 0.05$) and 1.2 cGy/h (Table 4.2) in comparison with the control, which would reduce the deficiency of proliferated cells. The stimulation needed to induce division in resting cells at low doses is well known (Luchnik 1958). The appearance of formerly resting cells newly proliferated in meristems can be the reason that the delay of time to fixing does not increase in cases of 0.3 and 1.2 cGy/h irradiation (Fig. 4.4a–c). The significant changes in survival, frequency of CCAs and the MI with 0.3 cGy/h irradiation suggest that the repair mechanisms were supplemented with additional regulation of cell numbers.

Testing of old seeds: In all groups of old seeds the average frequency of CCAs is higher than in the young seeds ($p < 0.05$), a finding that is in agreement with the results of previous investigations published in (Hanawalt 1987). The higher frequency of CCAs in aging seeds should reduce their viability (Navashin and Gerasimova 1935; Barton 1962), which could be the reason for the higher $(1 - S)$ values in all groups of old seeds (Table 4.2). In the "old" group control, both $(1 - S)$ and the frequency of CCAs were significantly elevated ($p < 0.05$) and the time to first fixation was delayed compared to the "young" group control (Fig. 4.3a). This delay can be understood in view of the elevated frequency of CCAs in rootlet meristems. In the old seeds control, neither the frequency of CCAs nor the MI value differed with time to fixing ($p > 0.025$). This had been anticipated because aging induces free radicals in seed cells (Pinzino et al. 1999) which can activate cell elimination mechanisms (Davis et al. 2001).

In seeds exposed at 0.3 and 1.2 cGy/h, neither the frequency of CCAs nor the MI values increased between the first-fixed and last-fixed fractions ($p > 0.05$) (Figs. 4.3b, c and 4.4b, c) as might have been expected from the elevated free radicals in all aging peas. In addition, it can be assumed that cell elimination was intensified at the 0.3 cGy/h dose rate because the $(1 - S)$ value was elevated ($p < 0.05$) and the average frequency of CCAs tended to (non-significantly) decrease compared with the old seeds control (Table 4.2). On the contrary, irradiation with 1.2 cGy/h increased the frequency of CCAs ($p < 0.05$), but not the $(1 - S)$ value ($p > 0.05$). Thus, one could suspect an impact of the changing predominance of cell elimination on CCAs emerging in the dose-rate interval 0.3–1.2 cGy/h. At these dose rates, the first fixations were earlier than in the control group of old seeds (Fig. 4.3a–c). The most probable mechanism of this phenomenon is stimulation of division in the less-damaged resting cells which renew the number of proliferating cells in meristem. This tendency was seen strongly at 0.3 cGyh dose rate ($p < 0.05$) (Fig. 4.4b, c; Table 4.2).

In the old seeds irradiated at 19.1 cGy/h, the time to fixing was delayed by up to 10 h compared with the controls, a delay that occurred in all three groups of exposed seeds (Fig. 4.3a, d). It would be expected that the frequency of CCAs and the MI value would be elevated in the last fixation, but they were not (Figs. 4.3d and 4.4d), perhaps due to aging of seeds. This result is in agreement with the elevated $(1 - S)$ value as in all other components of old seeds groups.

Aging modified the response of seeds upon irradiation. In any case, the old seeds demonstrated an increased average frequency of CCAs and decreased survival in comparison with the young ones. The 0.3 cGy/h dose-rate radiation induced the tendency to decrease both seed survival and frequency of CCAs, as well as to increase the MI as in the groups of young seeds.

Combined effect of irradiation and heat stresses: For this portion of the study, the young seeds were exposed both to elevated temperatures (30–32 °C) and radiation. In the heat-stressed control group, the frequency of CCAs increased with time to fixing ($p < 0.05$) in a manner similar to that observed for young seeds but the slope for the young + heat stressed seeds was reduced (Fig. 4.3a). It has been reported that an increase in temperature can induce apoptosis (Buzzard et al. 1998).

At 0.3 and 1.2 cGy/h, neither the frequency of CCAs nor the MI values increased significantly in the later fixed fractions, and the durations of seed germination were shortened (Figs. 4.3b, c and 4.4b, c). The average MI values did not increase (Table 4.2), giving indirect evidence that stimulation of proliferation, which contributes to seed survival, was low. These results are in agreement with the highest $(1 - S)$ value (32.5 and 62.5 %, respectively) ($p < 0.001$) in these groups of seeds.

In the case of irradiation with 19.1 cGy/h, the frequency of CCAs increased with time to fixing ($p < 0.001$) and the first fixation was delayed up to 10 h in comparison with the heat-stressed control (Fig. 4.4a, d). The MI value did not increase with time to fixing ($p > 0.05$), and it can be presumed that cells were eliminated. The averaged value of the MI was reduced in agreement with the high value of $(1 - S)$, 50 % (Table 4.2). Taken together, these results indicate that both CCAs and cell elimination were strongly stimulated.

The current observations over all seeds groups have shown that dose rates of 0.3, 1.2, and 19.1 cGy/h induced both cell elimination and CCAs that can be modified by aging and additional heat stress. As a rule, cell elimination dominated at the dose rate of 0.3 cGy/h and CCAs at 19.1 cGy/h. Irradiation with the dose rate of 0.3 cGy/h stimulated the MI and earlier seed germination; its period can be shorter (young and heat-stressed seeds) than in the control, while seed survival was decreased. Irradiation with 19.1 cGy/h delayed seed germination and did not decrease the frequency of CCAs while seed survival reduced significantly in aging and heat-stressed seeds.

4.2.3 Non-linearity Induced by Low-Dose Rates Irradiation. Effects of Aging and High Temperature

Dependence of the frequency of CCAs, theMI and survival of seeds on low-dose rate irradiation: Fig. 4.5 presents some of the experimental data as a function of dose rates. The $(1 - S)$ value is connected with the number of proliferated meristem cells which is calculated as the balance between elimination of cells and stimulation of proliferation. The frequency of CCAs (a, d, g) and the MI (b, e, h) values depend non-linearly on dose-rate (χ^2-criterion, $p < 0.05$) (Table 4.2), and that is the reason that $(1 - S)$ depends on the dose rate in a non-linear manner (χ^2-criterion, $p < 0.05$) (c, f, i).

Cell cycles can be perturbed by irradiation (Gudkov 1985; Burdon et al. 1989; Boei et al. 1996) and would include cell cycle delay (Gudkov 1985; Boei et al. 1996) as well as stimulation to proliferation (Luchnik 1958; Gudkov 1985). A cell cycle delay was shown at the dose rate of 19.1 cGy/h in all groups of seeds, and a stimulation to divide in resting cells was revealed at 0.3–1.2 cGy/h in groups of young (b) and old (e) seeds. This effect of mitotic stimulation was virtually

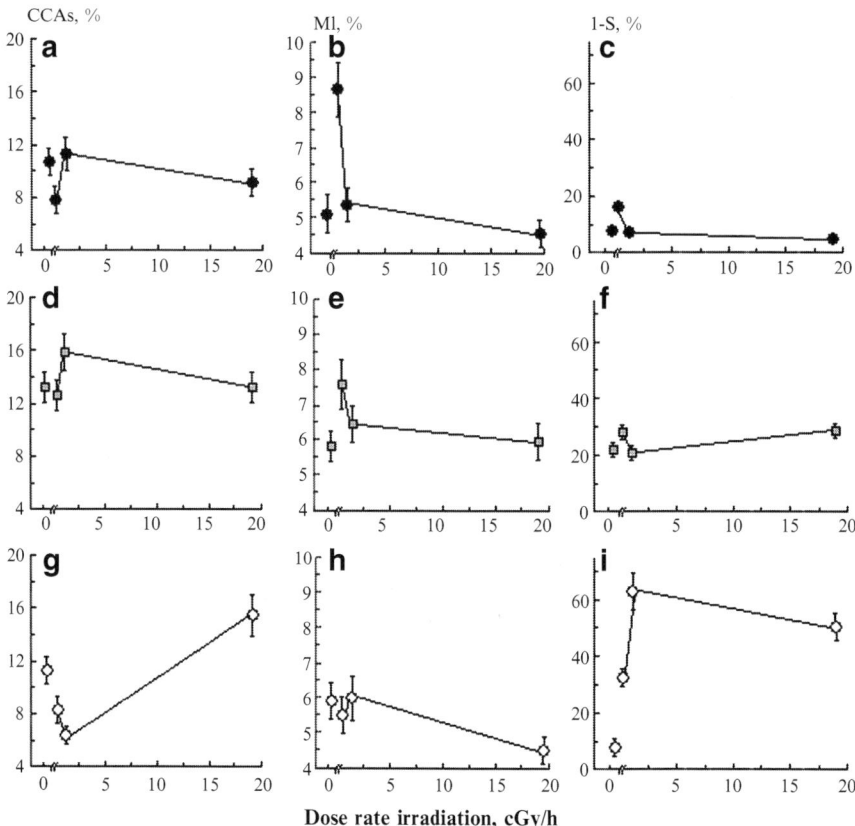

Fig. 4.5 Dependencies of the frequency of CCAs (**a, d, g**), MI (**b, e, h**), and (1 − S) (**c, f, i**) on dose rate. (**a–c**) young seeds, (**d–f**) old seeds, (**g–i**) young seeds with heat stress (heat stress duration 2 months). Values of the intact control seeds are separated from the irradiated ones by breaks

eliminated by the addition of heat stress (h). Perhaps stimulation compensates for cell elimination in seed groups exposed only to radiation at these dose rates while aging and exposure to heat appear to be more significant forces for decreasing viability.

We see that the group of old seeds (d–f) demonstrated the highest level of cells with abnormalities and its non-linearity is low in comparison with the characteristics of young seeds (a–c) population. Young seeds are distinguished from the old ones by strong cells' stimulation to proliferation (the heat-stressed seeds were not activated). The heat-stressed group of seeds (g–i) revealed strong non-linearity of the frequency of CCAs and (1 − S) value. We can suspect that dramatic differences are caused by the synergic effect of radiation and heat stresses.

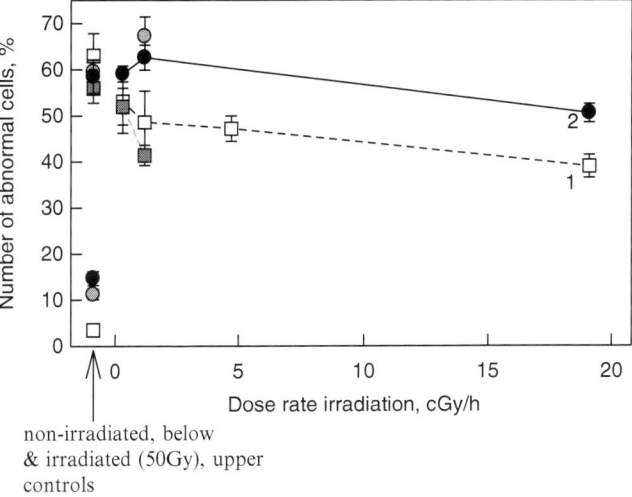

Fig. 4.6 Adaptive response of rootlet meristem cells of "young" seeds pre-irradiated at 7 cGy (*dashed line 1, white and grey squares*) and experienced combined effect of heat stress and low-dose irradiation (7 cGy) (*solid line 2, black and grey circles*) vs low dose rate irradiation. Acute γ-irradiation: 50 Gy. Control data are placed on the left of x-axis zero; non-irradiated control is *below* and the irradiated one at dose 50 Gy is *upper symbols*

4.2.4 Adaptive Response of Meristem Cells Induced by Low-Dose Rate Irradiation and Its Combination with High Temperature

These tests were carried out to study how the combination of radiation and heat stresses influences the induction of adaptive response.

The adaptive response of meristem cells was investigated for young and heat-stressed seeds pre-irradiated at 7 cGy and, for the second time, acutely irradiated at 50 Gy (Korogodina et al. 1998). The general tendency of the number of CCAs to decrease was observed in meristem of pre-irradiated seeds (Fig. 4.6). The principal difference has been revealed between reactions of meristem cells of the "young" seeds (the lower plot) and the "young seeds + heat stress" (the upper plot). For young seeds, the number of CCAs decreases with the dose-rate irradiation (0.3; 1.2; 19.1 cGy/h), whereas the appearance of CCAs tends to increase at irradiation with the dose rate of 0.3–1.2 cGy/h in heat-stressed seeds. The adaptive response for the heat-stressed group increases a little with dose-rate irradiation but it remains less in comparison with the response of cells in the young group of seeds. Experiments with both groups have been repeated and one of them on heat-stressed seeds is presented in Table 4.3.

The radiobiologists offered (Mothersill and Seymour 2004) that low-dose exposures cause removal of cells carrying potentially problematic lesions, prior to

Table 4.3 Number of meristem cells with abnormalities. Heat-stressed seeds were irradiated at low dose 7 cGy with dose rates 0.3; 1.2 and 19.1 cGy/h and acute dose 50 Gy

Dose rate irradiation, cGy/h	Number of seedlings	Number of ana-telophases	Number of cells with abnormalities,%
0	23	950	60.0 ± 7.4
0.3 cGy/h	35	2,943	58.5 ± 3.2
1.2 cGy/h	29	1,507	60.2 ± 5.1
19.1 cGy/h	24	1,838	49.2 ± 3.0

radiation exposure. Contrarily, synergism produces multiple lethal and sub-lethal lesions (Petin et al. 1999, 2002) which can be realized at additional acute radiation impact. It seems that synergic influence of radiation and heat stresses provides complete bad-repaired damages that decrease or "neutralize" the adaptive response. The summary is that the synergic effect of heat-radiation stress diminishes the adaptive response of cells especially at the dose-rate radiation of 0.3–1.2 cGy/h when the number of cells with abnormalities tends to increase.

At present, the bystander effect is documented at low doses in various investigations (Mothersill and Seymour 2000). The bystander effect suggests multiple appearances of cells with abnormalities, and we can check this hypothesis by statistical modelling.

4.3 Modelling of Appearance of Cells with Abnormalities

4.3.1 Appearance of Cells with Abnormalities in Seedlings' Meristem

The statistical analysis has revealed that seeds do not normally distribute on the number of CCAs, and the distribution displays a tail. It suggests the selection model (Florko and Korogodina 2007) of the appearance of CCAs based on the hypothesis of intercellular communication, see Sect. 3.4.1. The heterogeneity of the seed population suggests a division of seeds into resistant and sensitive subpopulations. The abnormal cells appear rarely and independently in the resistant subpopulation; they are accumulated in the seedling meristem without selection and are Poisson-distributed. The damaging process is more intensive and accompanied by selection in the sensitive seeds that can be described by the tail of their distribution on the number of cells with abnormalities (Florko and Korogodina 2007).

Analysis of the distribution structure of seeds on the number of abnormalities: The fitting analysis of the observed data verified that the seed populations consisted of two Poisson (^{s}P) and geometric (^{s}G) subpopulations (Florko and Korogodina

4.3 Modelling of Appearance of Cells with Abnormalities

Table 4.4 Mathematical modelling of appearance of cells with abnormalities

Dose rate, cGy/h	Number of seeds	Number of ana-telofases	Sample means		A value of a subpopulation	
			$^s mP$	$^s mG$	$^s N_P$	$^s N_G$
Young seeds						
0	60	2,202	1.6 ± 0.3	2.8 ± 0.4	0.34 ± 0.07	0.33 ± 0.07
0.3	60	2,294	1.1 ± 0.2	11.5 ± 4.3	0.60 ± 0.01	0.01 ± 0.10
1.2	60	2,774	1.5 ± 0.2	2.6 ± 0.6	0.50 ± 0.04	0.09 ± 0.09
19.1	60	2,144	0.8 ± 0.2	6.7 ± 1.2	0.58 ± 0.04	0.08 ± 0.09
Old seeds						
0	60	918	1.9 ± 0.3	0.0 ± 1.0	0.45 ± 0.00	0.00 ± 0.08
0.3	60	2,419	1.7 ± 0.2	0.0 ± 1.0	0.64 ± 0.00	0.00 ± 0.10
1.2	60	2,021	2.5 ± 0.3	10.1 ± 1.6	0.55 ± 0.03	0.06 ± 0.09
19.1	60	593	1.9 ± 0.3	0.6 ± 2.4	0.40 ± 0.00	0.00 ± 0.08
Young seeds with heat stress						
0	40	3,367	1.1 ± 0.2	13.3 ± 1.0	0.59 ± 0.08	0.31 ± 0.12
0.3	40	2,560	1.1 ± 0.2	∞	0.53 ± 0.05	0.11 ± 0.11
1.2	40	1,235	1.1 ± 0.3	0.0 ± 0.2	0.31 ± 0.03	0.05 ± 0.08
19.1	40	1,088	1.1 ± 0.3	7.3 ± 0.8	0.22 ± 0.08	0.28 ± 0.07

2007) (see Sects. 8.2.1 and 8.2.2). The modelling parameters $^s mP$ and $^s mG$ (the sample means) as well as $^s N_P$ and $^s N_G$ (the relative values) are shown in Table 4.4.

The values of the Poisson ($^s N_P$) and geometric ($^s N_G$) distributions were calculated as the ratios of a Poisson or geometric component to the whole tested seed population. In the dose-rate interval of 0.3–1.2 cGy/h, the $^s mG$ values tended to increase ($p < 0.05$), whereas the $^s mP$ ones did not change significantly ($p > 0.05$) in comparison with the control. In the groups of young and heat-stressed seeds, the $^s N_G$ values decreased in the dose-rate interval of 0.3–1.2 cGy/h (in the old seeds group, aging decreased $^s N_G$ too strongly and the statistical analysis is invalid). The $^s N_G$ quota decreased more than the $^s N_P$ quota ($p < 0.05$). As a rule, the parameters of the $^s P$ and $^s G$ subpopulations showed different behavior in the interval of 0.3–19.1 cGy/h for all groups of seeds (heterogeneity criterion (Glotov et al. 1982), $p < 0.05$).

We can assume that the bystander effect plays an important role in all the groups irradiated in the dose-rate interval 0.3–1.2 cGy/h especially in heat-stressed and old seeds. The quota of dead seedlings is significant even in the control group of old seeds having experienced oxidative stress (Pinzino et al. 1999; Davis et al. 2001).

Relation between radiation-induced processes: In young seeds, the value of $(1-S)$ correlated strongly with the MI ($|R|_{1-S.MI|CA} = 0.83$), and not with the frequency of CCAs (Table 4.5). This indicates that in young seed populations stimulation of proliferation is the general mechanism to regulate survival of the seeds irradiated with 0.3–19.1 cGy/h.

In old seeds, there is a correlation between $(1-S)$ and the frequency of CCAs ($|R| = 0.66$) that is due to the connection of seed survival with the N_G subpopulation ($|R|_{1-S.P|G} = 0.03$ and $|R|_{1-S.G|P} = 0.65$). The frequency of CCAs

Table 4.5 Partial correlations between the biological parameters, values Poisson (^sN_P is designated by P) and geometric (^sN_G is designated by G) distributions, and dose-rate irradiation

Coefficient	Young	Old	Heat stress
$\|R\|_{1-S,\,CCA\|MI}$	0.28	0.66	0.001
$\|R\|_{1-S,\,MI\|CCA}$	0.83	0.21	0.13
$\|R\|_{G,\,P\|MI}$	0.95	0.18	0.16
$\|R\|_{1-S,\,P\|G}$	0.21	0.03	0.99
$\|R\|_{1-S,\,G\|P}$	0.31	0.65	0.98
$\|R\|_{CCA,\,P\|MI}$	0.37	0.69	0.42
$\|R\|_{CCA,\,G\|MI}$	0.19	0.99	0.11
$\|R\|_{CCA,\,D\|MI}$	0.06	0.01	0.81
$\|R\|_{MI,\,D\|CCA}$	0.34	0.46	0.26
$\|R\|_{1-S,\,D\|MI}$	0.49	0.45	0.47
$\|R\|_{1-S,\,D\|CCA}$	0.53	0.61	0.39

also correlates significantly with ^sN_G ($p < 0.05$) ($\|R\|_{CCA,\,P\|MI} = 0.69$; $\|R\|_{CCA,\,G\|MI} = 0.99$). In addition, these values do not correlate with the radiation dose rate (Table 4.5). The ^sG subpopulation is considered to play an important role in aging seeds because the non-surviving fraction increases primarily as a result of diminution of ^sG seeds, and the frequency of CCAs also increases with the sample mean of the geometric distribution.

The combined effect of irradiation and heat stresses disturbed the correlations of $(1 - S)$ with both the frequency of CCAs ($|R| = 0.001$) and the MI value ($|R| = 0.13$). A strong correlation is observed between $(1 - S)$ and ^sN_P ($|R|_{1-S,\,P|G} = 0.98$) as well as ^sN_G ($|R|_{1-S,\,G|P} = 0.99$). Irradiation plays an important part in the case of heat-stressed seeds because the frequency of CCAs is related to the dose rate ($|R|_{CCA,D|MI} = 0.81$).

Irradiation with 0.3–19.1 cGy/h induces two regulatory mechanisms aimed at seed survival, stimulation of proliferation and the bystander effect. Stimulation of proliferation is sufficient to regulate viability in young seeds. In aging seeds, high frequency of CCAs is independent of the dose rate and is caused by sensitivity of cells in the ^sG subpopulation. A combination of irradiation and heat stresses induces another response of seeds. The frequency of CCAs depends on dose rate, and seed survival correlates with both P- and G-mechanisms of the appearance of CCAs. It should be noted that bystander effects lead to both enhanced frequency of CCAs and cell elimination. In the current studies, the cell elimination was observed mainly at 0.3 cGy/h and enhanced frequency of CCAs mainly occurred at 19.1 cGy/h. This summary is in agreement with the conclusion of C. Mothersill et al. (2000a) that the phenomenon of instability consists of the complete processes of mutation and lethal events which do not correlate.

The effect of low-dose rates was apparent not only as CCAs but also as increased values of $(1 - S)$ and MI, which require additional regulatory mechanisms. These are the bystander effect and stimulation of proliferation (Table 4.5). To conclude: the frequency of CCAs on average can't be sufficient criterion to define stress conditions because other mechanisms regulate seed viability and variability.

4.3 Modelling of Appearance of Cells with Abnormalities

Mechanisms that regulate seed viability and variability: The non-surviving fraction of seeds $(1 - S)$ was increased by even a single oxidizing factor but especially (up to 62 %) under the combined effect of multiple oxidizing factors. In all groups of seeds, $(1 - S)$ was non-linear at the interval of 0–19.1 cGy/h irradiation and increased significantly at a dose rate of 0.3 cGy/h (Table 4.2). For young seeds, the values of $(1 - S)$ correlated strongly with MI ($|R|_{1-S,\,MI|CCA} = 0.83$) while the frequency of CCAs did not ($|R|_{1-S,\,CCA|MI} = 0.28$) (Table 4.5). These results suggest that in young seed populations stimulation of proliferation is the general mechanism to regulate survival of seeds irradiated with 0.3–19.1 cGy/h.

For old seeds, the effects of irradiation were less pronounced on cell survival (30 % increase at 0.3 cGy/h compared with 100 % at the same dose rate for young seeds), but the impact of aging was dramatic with 2.5–5-fold increases in $(1 - S)$ (Fig. 4.6). In aging seeds, the frequency of CCAs and non-survival were correlated ($|R|_{1-S,\,CCA|MI} = 0.66$), and both values were independent of dose rate (Table 4.5). In this case high frequency of CCAs was specified by sensitivity of G- cells, and the correlation between the MI and $(1-S)$ was low ($|R|_{1-S,\,MI|CCA} = 0.21$) (Table 4.5).

The greatest mortality was observed in young + heat stressed seeds. When compared with the young seed group, the significantly increased $(1 - S)$ values indicate that heat stress produces a more significant impact on seeds than aging does (Table 4.2). The combined effect of irradiation and heat stresses disturbed the correlations of $(1 - S)$ with both the frequency of CCAs ($|R| = 0.001$) and the MI value ($|R| = 0.13$). In the heat-stressed group the frequency of CCAs depended on dose rate ($|R|$ ~$0.98 \div 0.99$) and seed survival was correlated with both P- and G- mechanisms of the appearance of CCAs (Table 4.5).

The MI increased significantly at 0.3 cGy/h for young and old seeds but not for young + heat stressed seeds. One can view the combined CCA and MI data in Table 4.2 as indicating that young and old seeds exposed to the lowest dose experienced a mitotic stimulus. Perhaps stimulation compensates for cell elimination in the groups exposed only to irradiation with tested dose rates (non-surviving fraction of young seeds increased up to 16 % only), while aging and exposure to heat appear to be more significant forces for decreasing viability (non-surviving fractions of old and heat-stressed seeds increased up to 28–62 %).

In all groups the frequency of CCAs decreased significantly at the lowest dose rate (0.3 cGy/h) ($p < 0.05$) that showed cell elimination. The number of CCAs was not normally distributed, but displayed tails. This result suggests that some CCAs were enhanced and correlated. The frequency of CCAs correlated with $(1 - S)$ in old seeds ($|R|_{1-S,\,CCA|MI} = 0.66$) (Table 4.5) but not in both young and heat-stressed seeds. Therefore, the frequency of CCAs cannot be sufficient criterion for the viability of plant population under radiation stress.

The simulation of the CCAs justifies the hypothesis that there are two subpopulations of seeds. In the first subpopulation, the CCAs appear independently and are Poisson-distributed. In the second one, the appearances of CCAs are correlated. The simulation supported both the Poisson and geometric mechanisms in meristems which contribute to seed survival (Table 4.5). In the dose-rate interval of 0.3–1.2 cGy/h, the *mG* values tended to increase ($p < 0.05$), whereas *mP* ones did not change significantly (p > 0.05) in comparison with the control. The (1 − S) value increased due to the elimination of cells and the failure of seeds to germinate in the geometric (particularly the Poisson) subpopulation (Table 4.4).

It can be suggested that the two mechanisms induced by stress conditions regulate the number of proliferating and abnormal cells. These are the bystander effect accompanied by selection, and stimulation of resting cells to divide. It seems that stress is an instrument to adapt populations in their ecological niches. Adaptation includes increasing variability (quickly elevating the sample mean mG) and dramatic decreasing of the number of germinated seeds; a new genotype of the surviving seeds is supported by repair mechanisms and stimulation of resting cells to divide. Stress-induced adaptation processes result in significant decreasing viability in the cases of aging and exposure to heat.

4.3.2 Chromosomal Instability

We shall examine below regularities of the genomic instability and accumulation of chromosomal aberrations by statistical modelling (see model in Sect. 3.4.2, and approximations in Sect. 8.2.3) and their connection with intercellular processes.

4.3.3 Multiple Appearances of Abnormal Chromosomes in Meristem Cells. Modelling of Chromosomal Instability

Adaptive processes in intact cells: Table 4.6 shows that the distribution of cells on the number of CAs relates to the geometric law (^cG) in the control group. It assumes that correlations between the appearance of CAs originated before "success" (Feller 1957), which could be considered in this case as the adaption of the genome to conditions. This suggests an adaptive process in intact cells: a DNA-damaging process coupled with selection which checks fitness of cells. It is true that an additional factor influences a sensitive subpopulation of cells in the first place, and its distribution on the number of CAs could be geometric (^cG) or Poisson (^cP). Thus, experimental data could be compounded of ^cG and ^cG (or ^cP) that correspond to distributions of resistant ($^cG1$) and sensitive ($^cG2$ or ^cP) fractions of cells. In the cases of aging and heat-stressed seeds, a combination of $^cG1$ and ^cP distributions is observed (Table 4.6). This is expected because high temperature and aging are strong mutagenic factors (Rainwater et al. 1996; Pinzino et al. 1999).

Distribution of cells in root meristem of irradiated seeds on the number of CAs: The modelling has shown two ^cG distributions in young seeds irradiated at 0.3 cGy/h (Table 4.7). It can be assumed that the second distribution $^cG2$ reflects additional damaging processes in the sensitive subpopulation. At 1.2 and 19.1 cGy/h, the ^cP distribution gradually displaces $^cG2$ distributions. Analyses of the combined effect

4.3 Modelling of Appearance of Cells with Abnormalities

Table 4.6 Statistical modelling of the appearance of CAs in root meristem cells of non-irradiated pea seeds

Year	CA frequency	Number of ana-telophases	Number of cells with CA number				
			0	1	2	3	≥ 4
Young							
1997	0.17	3,443	2,948	426	55	11	2
1997	0.12	2,199	1,931	238	25	5	-
1996	0.08	1,097	1,022	67	5	2	1
Heat stress							
1996	0.11	4,020	3,663	274	68	11	4
Old							
1996	0.10	906	823	77	4	1	1

The standard error of the frequency of CAs is ≤ 0.01. The fitting efficiency was assessed by the T (Geras'kin and Sarapultsev 1993) and $\chi^2 (p < 0.05)$ (Feller 1957) criteria

Table 4.7 Statistical modelling of the appearance of CAs in rootlet meristem cells of pea seeds irradiated with 7 cGy at low-dose rate or without irradiation

Dose rate, cGy/h	Number of studied ana-telophases	CA frequency	Number of cells with number of CAs				
			0	1	2	3	≥ 4
Young							
0	2,199	0.12	1,931	238	25	5	
0.3	2,139	0.08	1,971	145	19	4	1
1.2	1,843	0.07	1,637	191	13	2	
19.1	1,965	0.08	1,788	162	10	5	
Heat stress							
0	4,020	0.11	3,663	274	68	11	4
0.3	2,723	0.09	2,517	172	26	8	
1.2	1,380	0.08	1,296	68	9	6	1
19.1	1,274	0.18	1,092	143	29	6	4
Old							
0	906	0.10	823	77	4	1	1
0.3	2378	0.13	2,109	236	27	5	1
1.2	1,977	0.15	1,729	207	38	3	
19.1	586	0.11	527	53	5	1	

The standard error of CA frequency is ≤ 0.01. The fitting efficiency was assessed by the T (Geras'kin and Sarapultsev 1993) and $\chi^2 (p < 0.05)$ (Feller 1957) criteria

of radiation and high temperature have revealed domination of the G regularities of the appearance of CAs (heat stress) and addition of a P component in the case of aging (Table 4.7).

General characteristics and modelling values of multiple appearances of abnormal cells in meristem and abnormal chromosomes in cells: Table 4.8 presents general characteristics and modelling values of the appearance of both CCAs in meristem and CAs in cells. In the young seeds, a negligible value of seeds' G

Table 4.8 Seed nor-viabilities (1 − S), frequencies of both CCAs (R1) in seedling meristems and CAs in meristem cells (R2), parameters of the distributions of seeds on the number of cells with abnormalities as well as distributions of cells on the number of CAs

Dose rate	1 − S, %	R1, %	Distributions of seeds on the number of cells with abnormalities				R2****, %	Distributions of cells on the number of CAs					
			^sN_P	^smP	^sN_G	^smG		$^cN_{G1}$	$^cmG1$	$^cN_{G2}$	$^cmG2$	^cN_P	^cmP
Young													
0	8.0	10.7	0.33	1.6	0.34	2.8	0.12	1.00	0.88				
0.3	**16.0***	**7.8***	0.60	1.1	**0.01**	**11.5**	0.08	0.93	0.94	**0.07**	0.67		
1.2	7.3	11.4	0.50	1.5	0.09	2.6	0.07	0.26	0.87			0.74	0.06
19.1	5.5	9.1	0.58	0.8	0.08	6.7	0.08					1.00	0.09
Heat													
0	7.5	11.2	0.59	1.1	0.31	13.3	0.11	0.87	0.03				
0.3	**32.5***	**8.2***	0.53	1.1	0.11	20.0	0.09	0.70	0.02	0.30	0.25	0.13	0.66
1.2	**62.5***	**6.1***	0.31	1.1	**0.05**	**0.0**	0.08	0.94	0.04	**0.06**	0.64		
19.1	**50.0***	15.4*	0.22	1.1	0.28	7.3	0.18	0.85	0.11	0.15	0.60		
Old													
0	21.6	13.1	0.45	1.9	0.00	0.0	0.10	0.98	0.12	**0.01**	0.69	0.99	0.10
0.3	**28.1***	**12.5**	0.64	1.7	**0.00**	**0.0**	0.13	0.74	0.19	0.26	0.02		
1.2	20.7	15.8*	0.55	2.5	0.06	10.1	0.15	0.12	0.32				
19.1	**28.2***	**13.1**	0.40	1.9	**0.00**	**0.6**	0.11					**0.88**	**0.08**

The difference from the non-irradiated control: *$p < 0.05$; **$p < 0.001$; ***standard error ~8–12 %; ****standard error < 0.03. Standard errors of the parameters do not exceed 20–30 % (the sample means) and 10–15 % (the relative values). The fitting efficiency was assessed by the T (Geras'kin and Sarapultsev 1993) and χ^2 ($p < 0.05$) (Feller 1957) criteria

distribution (^sN_G) on the number of cells with abnormalities is coupled with a negligible value of cells' G2 ($^cN_{G2}$) distribution at 0.3 cG/h that is accompanied by decreasing frequencies of the corresponding abnormalities (R1, R2), and survival (Table 4.8). It is a reason to think that intensive radiation-induced bystander effects in meristem lead to intensive DNA damage in its cells that together decrease the survival of cells and seeds. Poisson distributions of seedlings and cells are observed at 1.2 and 19.1 cGy/h. At 19.1 cGy/h, the parameters ^smP (0.85 ± 0.20) and ^cmP (0.09 ± 0.02) have low values (Table 4.8).

In aged seeds, the negligible values of the ^sN_G and $^cN_{G2}$ distributions are observed at 0.3 cGy/h, that corresponds to the increased seed survival ($p < 0.05$). This increase could be due to stimulation of cell proliferation ($p < 0.05$) in the old seeds irradiated with 0.3 cGy/h dose rate (Korogodina et al. 2005). The ^sG component is also revealed in seeds irradiated at 1.2 cGy/h. The ^cP distribution is observed in the control and 19.1cGy/h groups, its sample mean being low (0.08–0.10 ± 0.03) (Table 4.8).

In the heat-stressed seeds, the lowest values of the ^sN_G and $^cN_{G2}$ distributions are observed at 1.2 cGy/h. The $^cG2$ distributions are revealed at 0.3 and 19.1 cGy/h; seed survival is decreased at all these dose rates, especially at 1.2 cGy/h. In the control group, the ^cP distribution of cells on the number of DNA damages is characterized by an increased sample mean (0.66 ± 0.12). All the above-mentioned could mean the increased late processes in all the groups of the seeds are induced by heat stress.

The comparison of the "old" and "heat" groups has shown that heat stress induces late processes accompanied by the selection stronger than the processes produced by aging.

4.3.4 Modification Effects of High Temperature and Aging

Synergic effect of radiation and heat stresses: A synergic coefficient is used to analyze quantitatively the combined effects induced by irradiation and heat. Petin et al. (1999) characterize the appearance of synergic lethal damages as $N_\Sigma = N_1 + N_2 + \min\{p_1N_1 + p_2N_2\}$, where N_1, N_2, and N_Σ are the numbers of the lethal damages induced by the first, second, and combined treatments; p_1 and p_2 are sub-lethal damages produced simultaneously. Then, the coefficient of synergism can be calculated as $K_{syn} = (N_1 + N_2 + \min\{p_1N_1 + p_2N_2\})/(N_1 + N_2)$.

Determination of the synergic coefficient can be used to its calculation. It is the ratio of a sum of probabilities of the appearance of abnormalities for each factor separately to their combination (see Sect. 8.3). According to these calculations, synergic characteristic of seeds' death (1 − S) increases strongly ($K_{syn} = 4$) (Fig. 4.7, pl. 3). Both numbers of combined DNA damages and combined damaged cells decrease ($K_{syn} < 1$) (Fig. 4.7, pl. 1, 2) in the "heat" groups that can be the result of elimination of "$^cG2$"-cells and "^sG"-seeds (Table 4.8). It is reasonable to conclude that adaptation to the combined effect of high temperature and low-dose irradiation is based on instability processes which lead to a dramatic synergistic death of cells and seeds.

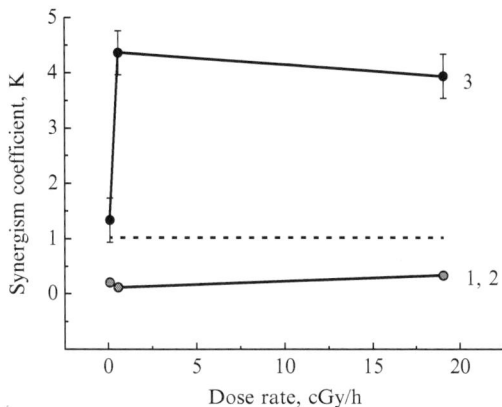

Fig. 4.7 Dependence of synergism coefficient on dose-rate irradiation. Synergism coefficients for the group of irradiated heat-stressed seeds: *1*, of the appearance of cells with CAs; *2*, of the appearance of CAs inside cells; *3*, of seeds' non-survival. The bottom of synergism is marked with a *dotted line*. Std. errors are shown

Low radiation effect on old seeds: The stimulation effect of irradiated plants was known many years ago (Gudkov 1985). A significant increasing of life span was shown on *Drosophila* (Moskalev et al. 2011). Examination of old seeds was performed to determine a hormetic effect. The hormetic coefficient was calculated as the ratio of a sum of probabilities of appearance of abnormalities (or death) caused by radiation and aging separately to their combination. These calculations (see Sect. 8.3) showed negligible decreasing of the frequency of CCAs (K_{horm} ~0.6), and the seeds' survival did not significantly change (K_{horm} ~0.75 ± 0.4 ÷ 1.03 ± 0.3).

The plant radiobiologists explain radiation stimulation by activation of proliferation (Gudkov 1985). We can try to clarify this hormetic phenomenon in terms of adaptation processes. Perhaps the resting pea cells were activated by the low-dose irradiation (Fig. 4.6), that renewed the proliferated pool which was free from the cells with abnormalities due to selection. So, the frequency of abnormal cells decreases and seeds' survival does not decrease.

4.3.5 Scheme of the Adaptation Process

The adaptive process has three components (Florko and Korogodina 2007; Korogodina and Florko 2007): primary radiation injury, which depends on the intensity of radiation (I_{rad}); late injury, which depends on the intensity of intercellular "bystander" (I_{byst}), and intracellular regulatory (I_{reg}) mechanisms; and selection which depends on repair systems and environmental conditions (Fig. 4.8).

These components can be presented by different combinations of Poisson and geometric laws where P statistics describes statistics of independent events (Florko and Korogodina 2007; Korogodina and Florko 2007) and G statistics characterizes statistics after selection (Orr 2006; Florko and Korogodina 2007; Korogodina and Florko 2007). The primary injury and late damage without selection follow the Poisson or binomial law (see Sects. 3.4.1 and 3.4.2). In both cases, the independently

4.3 Modelling of Appearance of Cells with Abnormalities

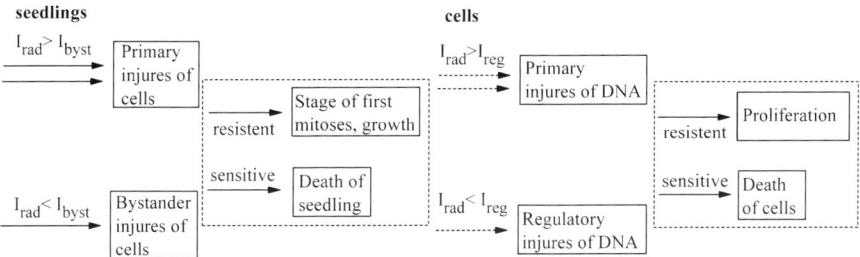

Fig. 4.8 Scheme of the process of seedlings' and cells' adaptation. Primary injury (I_{rad}) induces the intercellular bystander (I_{byst}) and intracellular regulatory (I_{reg}) processes. Primary injuries or late damages can be accumulated. As a result of selection, a stage of first mitoses is reached in resistant seedlings, and the sensitive ones die. Intracellular processes (primary or regulatory damaging of DNA) lead to proliferation or death of cells

occurring damages can be approximated as increasing linearly with time, that is, by the Poisson law with the increased sample mean (Florko and Korogodina 2007).

In practice, the observed P distribution is formed on the numbers of primary and late damages. If radiation intensity exceeds the rate of late process (Korogodina and Florko 2007) then the primary damages prevail although the P-sample mean is low (Florko and Korogodina 2007). For example, a dose rate of 19.1 cGy/h (time between two hits per cell = 5 s) induces hits in cells in the first two min with the averaged number of hits per cell ≈ 33. Table 4.8 shows that this case corresponds to the lowest sample mean of the $^s mP$ distribution of seedlings on the number of damaged cells (0.85 ± 0.20). Irradiation at dose rates of 0.3 and 1.2 cGy/h (time between two hits = 5 and 1.3 min, respectively) induces bystander effects which should prevail in these cases. The complete analysis of the pea data (Table 4.8) reveals the correspondence of low $^s mP$ values to low $^c mP$ values which couple with a high damage factor. The increased values of $^s mP$, $^c mP$ indicate late processes (for example, the control heat-stressed group, Table 4.8).

Thus, in experiments on pea seeds, the distribution of seedlings on the number of CCAs can be approximated by $^s P + {}^s G$, where the $^s P$ distribution relates to the subpopulation of seedlings in which the bystander effect is not accompanied by the death of plants, and $^s G$ describes the bystander damage accompanied by seedling selection in a more sensitive fraction (Florko and Korogodina 2007; Korogodina et al. 2010).[3] A distribution of cells on the number of CAs can be presented as $^c G1$, $^c G2$, and $^c P$ distributions and their combination (Tables 4.6 and 4.7) (Korogodina et al. 2010). The $^c P$ distributions mean primary DNA damage which, together with $^c G2$ distributions, describes the appearance of DNA damage in the sensitive fraction of cells. The appearance of the adaptive process in the resistant fraction of cells is described by the $^c G1$ distribution. When the primary exposure intensity exceeds intracellular regulatory effects, the Poisson law displaces the G-one (Korogodina et al. 2010).

---

[3] The cells stimulated to proliferate are involved in the distributions.

4.4 Summary

General regularities of the low-dose-rate radiation effects were determined on plant seeds in laboratory experiments. Three groups of seeds were tested: young, old, and heat stressed. Effects of aging and high temperature were studied because they are standard natural factors. Then, the statistical modelling was performed to consider some processes which are interesting from our point of view.

General characteristics: We can see that the radiation effect is different in the interval of 0.3–19.1 cGy/h dose-rate radiation. Irradiation with 19.1 cGy/h delays seeds germination. The lower intensities of 0.3 and 1.2 cGy/h can lead to decreasing of the frequency of CCAs and shortened period of germination (Korogodina et al. 2005).

The non-linearity of the frequency of CCAs, non-survival $(1 - S)$ and MI was revealed in the dose-rate interval of 0.3–1.2 cGy/h. Dramatic differences are observed in young and especially heat-stressed groups (the heat-stressed seed survival has fallen up to 30–40 %). In old seeds, the frequency of CCAs is always high, and their survival is low.

It was revealed that the synergic effect of heat-radiation stress diminishes the adaptive response of cells especially at the dose-rate radiation of 0.3–1.2 cGy/h when the number of cells with abnormalities tends to increase (Korogodina et al. 1998).

Statistical modelling of the appearance of CCAs: The experimental data were used to analyze the processes induced by radiation. The fitting has shown that the model "P + G" describes efficiently the distributions of seeds on the number of CCAs. We can be sure that the geometric law relates to the selection (Orr 2006) and the model "P + G" corresponds to the adaptation process (Florko and Korogodina 2007). The adaptation process has three components: primary injures, late damaging, and selection. They are more intensive in the sensitive subpopulation of seeds. We see that the late damaging and selection are more pronounced in the old and heat-stressed groups at irradiation of the 0.3–1.2 cGy/h dose rate because respective G-sample means have significantly increased (due to the bystander effect) or, on the contrary, decreased (due to selection) (Korogodina et al. 2005).

The correlation analysis has shown the relation between P-, G-parameters and biological values (Korogodina et al. 2005). We can describe the main regularities of viability and variability on the basis of these correlations. They are different for three groups of seeds. For young seeds, the stimulation of proliferation is the general mechanism to regulate the survival of seeds irradiated with 0.3–19.1 cGy/h. Aging is the main reason of the high frequency of CCAs and selection in old seeds. In the heat-stressed seeds, the frequency of CCAs is determined by the dose-rate irradiation, and dramatic mortality correlates with both P- and G-values that reflect the characteristics of the radiation intensity, bystander effect and repair intensity.

Multiple appearances of CAs: The analysis of the intracellular processes shows permanent adaptation process in intact cells (G1) (Korogodina et al. 2010).

Irradiation induces primary damages in sensitive cells that lead to more intensive instability and selection processes there (G2) or accumulation of primary damages (P) if primary intensity exceeds the late process. We see the relation between inter- and intracellular processes. The strong selection relates to the 0.3 and 1.2 cGy/h dose rates that can be described by two G-distributions on the number of CAs. The dose rate of 19.1 cGy/h induces dominant primary DNA damages in young and old seeds (we see P-law). The combined effect of radiation and heat stresses is always accompanied by the G-laws that mean instability and selection.

Modification effects of heat stress and aging: Calculations of the "combination coefficient" have demonstrated a strong synergic effect of radiation and heat stresses especially for seed mortality. The synergic coefficient for the frequency of abnormalities is less than unity because it indicates the deceptive "hormetic effect", which means only high G-mortality of seedlings with multiple numbers of abnormalities.

This study was performed to look for the hormetic effect in old seeds. The calculations revealed a tendency to decrease the frequency of abnormalities. Seed survival did not change significantly in comparison with those that were non-irradiated.

References

Atayan RR (1987) Interaction of factors modifying the radiosensitivity of dormant seeds: a review. Int J Radiat Biol Relat Stud Phys Chem Med 52:827–845

Barton L (1962) Seed preservation and longevity. Mark and Phyll Publ, London

Boei JJ, Vermeulen S, Natarajan AT (1996) Detection of chromosomal aberrations by fluorescence in situ hybridization in the first three postirradiation divisions of human lymphocytes. Mutat Res 349:127–135

Burdon RH, Gill V, Rice-Evans C (1989) Cell proliferation and oxidative stress. Free Radical Res Commun 7:149–159

Buzzard KA, Giaccia AJ, Killender M et al (1998) Heat shock protein 72 modulates pathways of stress-induced apoptosis. J Biol Chem 273:17147–17153

Davis W Jr, Ronai Z, Tew KD (2001) Cellular thiols and reactive oxygen species in drug-induced apoptosis. J Pharmacol Exp Theor 296:1–6

Dineva SB, Abramov VI, Shevchenko VA (1993) Genetic effects of lead nitrate on seeds of chronically irradiated populations of *Arabidopsis thaliana*. Genetika 29:1914–1920

Evseeva TI, Geras'kin SA (2001) Combined effect of radiation and non-radiation factors on Tradescancia. Ural Division of RAS, Ekaterinburg (Russian)

Feller W (1957) An introduction to probability theory and its applications. Wiley/Chapman & Hall, Limited, New York/London

Florko BV, Korogodina VL (2007) Analysis of the distribution structure as exemplified by one cytogenetic problem. PEPAN Lett 4:331–338

Geras'kin SA, Sarapultsev BI (1993) Automatic classification of biological objects by the radiation stability level. Autom Telemech 2:183 (Russian)

Glotov NV, Zhivotovskjy LA, Khovanov NV et al (1982) Biometrics. Leningrad State University, Leningrad (Russian)

Gudkov IN (1985) Cell mechanisms of postradiation repair in plants. Naukova Dumka, Kiev (Russian)

Hanawalt PS (1987) On the role of DNA damage and repair process in aging: evidence for and against. In: Werner HR, Butler RN, Sprott RL et al (eds) Modern biological theories of aging. Raven, New York, pp 183–198

Kim JK, Petin VG, Zhurakovskaya GP (2001) Exposure rate as a determinant of the synergistic interaction of heat combined with ionizing or ultraviolet radiation in cell killing. J Radiat Res (Tokyo) 42:361–369

Korogodina VL, Florko BV, Korogodin VI (2005) Variability of seed plant populations under oxidizing radiation and heat stresses in laboratory experiments. IEEE Trans Nucl Sci 52:1076–1083

Korogodina VL, Florko BV, Osipova LP (2010) Adaptation and radiation-induced chromosomal instability studied by statistical modeling. Open Evol J 4:12–22

Korogodina VL, Florko BV (2007) Evolution processes in populations of plantain, growing around the radiation sources: changes in plant genotypes resulting from bystander effects and chromosomal instability. In: Mothersill C, Seymour C, Mosse IB (eds) A challenge for the future. Springer, Dordrecht, pp 155–170

Korogodina VL, Panteleeva A, Ganicheva I et al (1998) Influence of the low gamma-irradiation dose rate on mitosis and adaptive response in meristem cells of pea seedlings. Radiats Biol Radioecol 38:643–649 (Russian)

Luchnik NV (1958) Influence of low dose irradiation on mitosis in pea. Newslett Ural Department Moscow Soc Nat Investig 1:37–49 (Russian)

MacCarrone M, Van Zadelhoff G, Veldink GA et al (2000) Early activation of lipoxygenase in lentil (*Lens culinaris*) root protoplasts by oxidative stress induces programmed cell death. Eur J Biochem 267:5078–5084

Moskalev AA, Plyusnina EN, Shaposhnikov MV (2011) Radiation hormesis and radioadaptive response in *Drosophila melanogaster* flies with different genetic backgrounds: the role of cellular stress-resistance mechanisms. Biogerontology 12(3):253–263

Mothersill C, Kadhim MA, O'Reilly S et al (2000a) Dose- and time-response relationships for lethal mutations and chromosomal instability induced by ionizing radiation in an immortalized human keratinocyte cell line. Int J Radiat Biol 76:799–806

Mothersill C, Stamato TD, Perez ML et al (2000b) Involvement of energy metabolism in the production of 'bystander effects' by radiation. Br J Cancer 82:1740–1746

Mothersill C, Seymour CB (2000) Genomic instability, bystander effect and radiation risks: implications for development of protection strategies for man and environment. Radiats Biologija Radioecologija 40:615–620

Mothersill C, Seymour CB (2004) Radiation-induced bystander effects and adaptive responses – the Yin and Yang of low dose radiobiology? Mutat Res 568:121–128

Navashin M, Gerasimova E (1935) The origin and reasons of mutations. Biol J 4:593–642 (Russian)

Ohba K (1961) Radiation sensitivity of pine seeds of different water content. Hereditas 47:283–294

Orr HA (2006) The distribution of fitness effects among beneficial mutations in Fisher's geometric model of adaptation. J Theor Biol 238:279–285

Petin VG, Kim JK, Zhurakovskaya GP et al (2002) Some general regularities of synergistic interaction of hyperthermia with various physical and chemical inactivating agents. Int J Hyperthermia 18(1):40–49

Petin VG, Zhurakovskaia GP, Pantiukhina AG et al (1999) Small doses and synergistic interaction of environmental factors. Radiats Biol Radioecol 39(1):113–126 (Russian)

Pinzino C, Capocchi A, Galleschi L et al (1999) Aging, free radicals, and antioxidants in wheat seeds. J Agric Food Chem 47:1333–1339

Preobrazhenskaya E (1971) Radioresistance of plant seeds. Atomizdat, Moscow (Russian)

Rainwater DT, Gossett DR, Millhollon EP et al (1996) The relationship between yield and the antioxidant defense system in tomatoes grown under heat stress. Free Radic Res 25:421–435

Seymour CB, Mothersill C (2000) Relative contribution of bystander and targeted cell killing to the low-dose region of the radiation dose–response curve. Radiat Res 153:508–511

Chapter 5
Adaptation and Genetic Instability in Ecology. Study of the Influence of Nuclear Station Fallout on Plant Populations

Abstract Here, the ecological investigations which were performed by statistical modelling are presented. The aim of these investigations was to study the effects of the fallout of operating nuclear station which do not exceed the background (the annual γ-radiation dose rates are ≈0.10–0.15 μSv/h) on plantain populations growing in the 30-km zone of the station. The statistical modelling was performed to study cells' and chromosomes' instability, cells' proliferation in root meristem of seedlings, selection processes, and their dependence on radiation fallout and seeds' sensitivity. This approach divides the tested population into resistant and sensitive fractions which are Poisson and geometric distributed on the number of abnormalities. This finding allowed us (1) to identify the radiation-induced effect and to reject the hypothesis that its consequences were caused by chemical pollution; (2) to estimate risks of instability and selection as well as stimulation of cells' proliferation which are components of the adaptation processes; (3) to study their regularities for both resistant and sensitive subpopulations; and (4) to demonstrate that risks of adaptation processes dramatically increase with additional synergic factors, such as in a hot summer when survival of seeds falls to 20–30 %. We conclude that nuclear station fallout influences significantly the seeds' survival and chromosomal instability in meristem of seedlings especially in hot summer. The radiation-induced processes lead to changes of the previous genotype on the adaptive one. The other conclusion is decreased numbers of some species right up to their disappearance.

Keywords Operating nuclear station fallout • Statistical modelling • Natural plantain populations • Seeds • Radiation and high temperature synergism • Chromosomal abnormalities • Stimulation of proliferation • Resistance and sensitivity • Risk of instability

5.1 A View on Natural Communities with Experience in Radiation Impact

It is known that radiation impact leads to genetic consequences for human and natural populations. It is shown that irradiation which does not exceed the limits of ecological niches can influence living organisms. The main consequences for nature could be considered with the example of the Chernobyl accident (1986) because scientists have collected a lot of data related to this radiation accident for different species.

5.1.1 Consequences of Radiation Accidents

The Chernobyl accident fallout immediately annihilated the forest in the 30-km zone ("red forest") (Yablokov et al. 2009). Later, the main injuries to plants were caused by incorporated radionuclides in the soil. Plants were transformed into radionuclide accumulators in the polluted areas (Geras'kin et al. 2008). The Chernobyl impact resulted in radiomorphoses and tumors because the radiation disturbed morphogenesis and provoked tumorogenesis (Grodzinsky et al. 1991; Grodzinsky 1999). The number of pollen abnormalities in the polluted territory has remained high for a generation (Kovalchuk et al. 2000).

Genetic consequences: Chernobyl radiation immediately elevated the frequency of plant mutations, which remained high for some years, and the level of mutation frequency was correlated with the level of radiation pollution (Shevchenko et al. 1996; Geras'kin et al. 2008; Grodzinsky 2009). The mutation frequency can increase, normalize, or decrease in a generation for different plant species (Grodzinsky et al. 1996). The elevation of mitotic activity (MA) and disturbance of mitosis was determined for different species (Artyukhov et al. 2004). Survival of seeds (*Crepis tectorum*) didn't exceed 40–50 % in the Chernobyl 30-km zone (Shevchenko et al. 1995).[1] Some scientists demonstrated the non-linear dependence of cytogenetic values (Shevchenko 1997; Geras'kin et al. 2008).

The radiation resulted in reversion of human-cultivated types of plants and animals to their wild types (Glazko 2006). The induced atavistic features are typical for the extinct ancestor forms (Yablokov et al. 2009).

Romanovskaya et al. (1998) studied bacteria in the contaminated soil. The tested population of cellulolytic, nitrifying, and sulfate-reducing bacteria was found to be 1–2 orders of magnitude less than in the control soil that indicates the unfavorable effect of anthropogenic radiation on the abundance and diversity of soil bacteria. The most exposed phytocenoses and soil animals' communities have exhibited dose-dependent alterations in the species composition and reduction in biological diversity, although Geras'kin et al. (2008) did not observe any decrease in the number of small mammals or taxonomic diversity, even in the radioactive habitat.

---

[1] Usually survival of seeds in nature is ~80 % (Preobrazhenskaya 1971).

5.1.2 Radiation Sources Providing Low-Radiation Stress Comparable with Background

Nowadays, the argument is made that radiation comparable to the background influences both humans and nature. In this connection, we will consider sources of radiation stress and adaptation response on their impact.

At present, there are many radiation sources which provide low irradiation. It is difficult to find a site anywhere in the world which does not more or less experience some influence of anthropogenic radiation. There are territories which were radiation-polluted after nuclear tests and nuclear accidents. Mineral trades, radiochemical industries, and scientific and industrial reactors also contribute to this. One of the essential sources of low-radiation pollution is operating nuclear power plants (NPP). One could expect the influence of NPPs on the environment, although their fallout does not exceed the background. The operating standard Balakovo NPP was chosen to study the influence of its radiation fallout on the environment.

The Balakovo nuclear power plant: Balakovo NPP is located near the town of Balakovo, Saratov region, Russia, about 900 km to the southeast of Moscow. It consists of four operational reactors; the fifth unit is still under construction. The owner and operator of the NPP is Rosenergoatom. Balakovo NPP participates in the joint program between NPPs in Europe and Russia; since 1990, it has been in partnership with the Biblis NPP.

Ecological safety of the Balakovo NPP is standard. Uncontrolled influence of the radioactive materials of the NPP on the environment was excluded from the project. The only planned standardized source of the radiation influence on the environment is fallout through the ventilation tubing. Values of the average daily fallout are less than the maximum permissible dose by a factor of a hundred-thousand. The observation zone around the NPP is 30 km, the sanitary zone is 2.5–3.0 km. Contamination by radionuclides in the environment around the NPP is within the average values which are characteristic for the European part of Russia.

Nowadays there are many regions in Russia which are polluted both with nuclear and chemical industry wastes. The Balakovo area in the region of Saratov is a place where radionuclide contamination and chemical pollution are caused by benzopyrene, carbon bisulphide, heavy metals and others. In said area, there are NPPs, chemical fertilizer factories, produced from the apatite on the Kola Peninsula, and plenty of representation of the two largest enterprises in Russia: rubber and chemical fiber production.

5.2 Characteristics of Objects and Methods in the Ecological Studies

The ecological investigations were performed near the Balakovo NPP for 2 years (1998, 1999) by specialists from the Saratov State University (ecology, botany), and the Saratov and Balakovo Nature State committees (ecology). It allows one

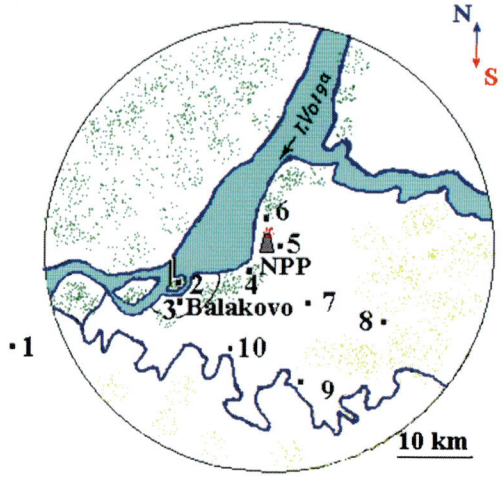

Fig. 5.1 Locations of the selected plantain populations in the vicinity of the Balakovo NPP: *2* (P2); *3* (P3); *4* (P4); *5* (P5); *6* (P6); *7* (P7); *8* (P8); *9* (P9); *10* (P10). Populations P1, P11 are located 80 km and 100 km from NPP, respectively; P12 is placed in the Moscow region

to choose some plantain populations in similar biogeocenoses (Korogodina et al. 2004). The radiation dosimetry, radiochemistry analysis of soil, cytogenetic analysis and mathematical modelling were performed at the Joint Institute for Nuclear Research (JINR).

It is known that reactive oxidative species (ROS) are involved in the stress-induced bystander mechanism (Averbeck 2010). The oxidants have an important role in plant gene expression and gene-product regulation (Foyer and Noctor 2009). Cellular redox homeostasis is considered to be an "integrator" of information from metabolism and the environment controlling the plant growth and acclimation responses. Low molecular antioxidants serve not only to limit the lifetime of the ROS signals but also to participate in an extensive range of other redox signaling and regulatory functions. The analysis of the antioxidant status (AOS) of seeds was performed at the Moscow Institute of Biochemical Physics of RAS (biochemistry). Results of these investigations were published in a series of papers (Burlakova et al. 1998; Korogodina et al. 2000, 2006, 2010a, b; Korogodina and Florko 2007).

Characteristics of sites and weather conditions: The sites were chosen near the NPP and in the JINR territory to investigate the possibility of radiation stress effects. One site was studied in the Chernobyl trace territory to compare the effects of low-dose rate irradiation caused by atmospheric impact and chronic soil pollution.

Two sources of radioactivity were placed in the 30-km zone of the NPP, which can influence the plantain populations: the atomic station (P2–P6 sites) and the phosphogypsum dump (P7–P10 sites) (Fig. 5.1). The sites were chosen by use of a wind rose. The NPP atmospheric fallout (The Ministry of Atomic Power of the RF et al. 1998a) influence the better part of populations P2–P6 resulting from the direction of the winds in summer. Perhaps populations P7–P10 experience the effects of the phosphogypsum dump in this area (State Committee on the Environment Protection in the Saratov Region 2000). Site P1 is on the left bank of the Volga River (~80 km from NPP). Population P11 was chosen on the right

5.2 Characteristics of Objects and Methods in the Ecological Studies

Table 5.1 Soil contamination of radionuclides ^{137}C and ^{40}K (Bq/kg)

Radionuclide	Year	P1	P2	P3	P4	P5	P6	P7	P8	P9	P10	P11	P12
^{137}C	1998	9	32	–	5	–	–	10	8	8	6	39	9
	1999	5	4	3	5	5	5	–	5	–	–	15	10
^{40}K	1998	360	330	–	550	–	–	470	450	460	700	450	580
	1999	440	340	400	410	230	500	–	410	–	–	320	600

bank of the Volga (100 km from the NPP) on the Chernobyl radioactivity-deposition track with well-characterized ^{137}Cs soil contaminations (average concentration ~30 Bq/kg, (European Commission 1998)). Site P12 was selected in the Moscow region within the JINR territory so both the radiation exposures from the Institute facilities and the soil pollution would be known. A description of the biogeocenoses is published in (Korogodina et al. 2000).

For the Saratov region, the annual rainfall near the Volga is 1.5 times higher than in the steppe, therefore the microclimate of the P1–P6 sites is damper than that of the P7–P10 (State Committee on the Environment Protection in the Saratov Region 2000). In 1999, the high summertime temperatures in the European part of Russia averaged 3–4 °C above normal (extreme summer temperatures) (State Committee on the Environment Protection in the Saratov Region 2000).

Determination of radioactivity and accumulative doses: For sites within a 100 km radius of the NPP, the annual γ-radiation dose rates (DR) and ^{137}Cs soil concentrations (C_{Cs}) varied in the ranges ~0.10–0.15 μSv/h and ~5–10 Bq/kg as reported in independent radiological surveys (The Ministry of Atomic Power of RF et al. 1998b, a; State Committee on the Environment Protection in the Saratov Region 2000). In site P11, DR is ~0.10–0.15 μSv/h (State Committee on the Environment Protection in the Saratov Region 2000) and C_{Cs} is 30 Bq/kg (European Commission 1998). DR is ~0.10–0.12 μSv/h and C_{Cs} is ~5–10 Bq/kg in site P12 (Dubna) (Alenitskaja et al. 2004). These values (excluding the concentrations in site P11) do not exceed the average radiation values over the Saratov and Moscow regions (State Committee on the Environment Protection in the Saratov Region 2000; Alenitskaja et al. 2004; Zykova et al. 1995).

The upper 10–12 cm of soil in the tested sites was examined. Measurements were carried out using the low-background γ-spectrometers with a NaI(Tl) crystal as well as the Ge one, which were described in (Alenitskaja et al. 2004; Bamblevskij 1979). The errors of detection efficiency of γ-quanta did not exceed 7 %. Total errors of radioactivity determination for different isotopes were 20–40 %. The artificial isotope ^{137}Cs soil contamination did not differ significantly in 1998 and 1999 (Table 5.1), although the fluctuations could usually be observed at the same site (Alenitskaja et al. 2004). The data on C_{Cs} is in agreement with the published values (State Committee on the Environment Protection in the Saratov Region 2000; Alenitskaja et al. 2004; Zykova et al. 1995) and do not correlate with the NPP fallout.

The accumulated doses were calculated by using the Brian-Amiro model (Amiro 1992) and estimated as ~1–3 cGy for plantain seeds P1–P6 using the published transfer factors (Alenitskaja et al. 2004). The secondary wind rising in the P7–P10

Table 5.2 Relative daily Balakovo NPP fallout (radioisotopes Kr, Xe, I) dose rates DR, experienced by populations, which were calculated in the ratio to the dose in site P1

Populations	Relative dose rate (DR)	Intensity of γ-quanta per cell nucleus per min $\times 10^{-7}$	Intensity of γ-quanta per cell nucleus per 3 months
P1	1	$1.9 \cdot 10^{-4}$	$2.5 \cdot 10^{-6}$
P2	80	$1.7 \cdot 10^{-2}$	$2.2 \cdot 10^{-4}$
P3	80	$1.7 \cdot 10^{-2}$	$2.2 \cdot 10^{-4}$
P4	560	0.11	$1.4 \cdot 10^{-3}$
P5	5,700	65	0.85
P6	1,350	0.26	$3.4 \cdot 10^{-3}$
P7	340	$6.7 \cdot 10^{-2}$	$8.6 \cdot 10^{-4}$

populations was accounted by means of the methods described in (Gusev and Beljaev 1986). The results did not differ significantly from 1998 to 1999 (for accumulated radiation doses and soil concentrations) in the tested sites.

Calculation of NPP fallout and JINR accelerators' irradiation of seeds: Plantain seeds were not irradiated in the lab, but experienced the NPP fallout irradiation in nature (annual fallout on isotopes: Kr ~2.5 TBq; Xe~2.5 TBq, and I ~4.4 TBq (The Ministry of Atomic Power of RF et al. 1998b), the dose rates are controlled by the NPP administration).

Distribution of particulate emissions and gases were estimated according to the Smith-Hosker model (Hosker 1974) based on NPP characteristics (The Ministry of Atomic Power of RF et al. 1998b) and winds in summer near the ground in the NPP region (State Committee on the Environment Protection in the Saratov Region 2000). The isotope fallout result in γ-irradiation mainly (mean energy ~1.1 MeV/γ-quanta (Ivanov et al. 1986)). The relative DR values were calculated in the ratio to the dose in site P1 (Table 5.2). The relative DR value in the P7 site is higher than in the P2, P3, P8–P10 sites due to a short half-life of I isotopes, which do not reach populations P2, P3, P8–P10. This fallout irradiation is not chronic and depends on location of populations. The intensity of irradiation is shown in Table 5.2. The γ-quanta' LET-dependence on their energy (Ivanov et al. 1986) and the NPP characteristics (The Ministry of Atomic Power of RF et al. 1998b) were used to calculate a mean γ-quanta energy deposition per plant cell nucleus, which is 1.4 KeV.

The expected irradiation dose was calculated after the JINR facility's operation in the P12 site. In 1998, the calculated neutron dose level was 1 mSv (for 2 months), and the neutron dose rate level was 0.8 μSv/h (the neutron background dose rate ~9.3 nSv/h (Wiegel et al. 2002)); the radiation level over all the particles exceeded the background ~twofold for the dose and ~eightfold for the dose rate. In 1999, the neutron irradiation did not increase. The averaged neutron energy was 5.5 MeV.

Chemical pollutions: Soil samples were examined by X-ray fluorescence and gamma activation at the JINR by the method described in (Maslov et al. 1995). Chemical pollutions of the atmosphere and soil were published in (State Committee on the Environment Protection in the Saratov Region 2000; Korogodina et al. 2000). The contamination of the chemical pollutions was less in 1999 than in 1998. The correlative analysis was used to study the dependence of both biological and modelling values on the NPP fallout (see Sect. 8.5). The analysis has shown that neither the NPP fallout nor biological values correlate with the concentration of chemical pollution in this region (Korogodina et al. 2010a). It is supposed that chemical pollution cannot be a reason for the radiobiological regularities and also cannot imitate them.

Seeds: Plantain seeds (*Plantago major*) were used in the natural experiment. For this kind of seed, the reported quasi-threshold radiation dose, which corresponds to the inflection of the survival curve from a shoulder to mid-lethal doses, is 10–20 Gy (Preobrazhenskaya 1971). The plantain populations were located at sites within 80 km of the Balakovo NPP and in the Chernobyl trace area (Saratov region), as well as in the JINR territory (Moscow region). In 1999, temperatures during daylight hours reached 30–32 °C in the Moscow region and 38–40 °C in the Saratov territory (it is the extreme t° for these provinces), and the seeds experienced elevated temperatures during the maturation period in nature (The Ministry of Atomic Power of the RF et al. 1998a; State Committee on the Environment Protection in the Saratov Region 2000). The plantain populations were chosen in similar biotopes (Korogodina et al. 2000), which are located along the reservoir banks. The seeds were collected from 20 to 30 plants at the end of August in 1998 and 1999. The seeds were refrigerated until the next April.

Cytogenetic analysis: Seeds of all populations were germinated until seedling roots reached the length which corresponds to the first mitoses (Korogodina et al. 2004, 2010a). Ana-telophases were scored for the cells containing chromosome bridges and acentric fragments. The MA of seedling meristem cells was scored as the number of ana-telophases in apical meristem.

Estimation of antioxidant status: Peeled seeds (1 g) were placed in 200 ml of water at 60 °C, allowed to cool to room temperature for 1 h, and filtered. Antioxidant activities were studied with a photochemiluminescence method (Lozovskaya and Sapezhinskii 1993). It was shown that the reciprocal relative intensity of chemiluminescence I_0/I (where I_0 is the control intensity and I is the intensity in the presence of seed infusion) could be satisfactorily described as a linear function of the added infusion. This dependence allowed one to find the amount of infusion, which inhibited chemiluminescence by 50 % ($C_{1/2}$). The reciprocal value of $C_{1/2}$, linearly proportional to the inhibiting effect, was adopted as a measure of AOS.

5.3 Analysis of Averaged Biological Values

AOS of plant seeds: Investigation of AOS of plantain seeds collected in testing sites (Korogodina et al. 2000) have shown differences in their antioxidant activity (Table 5.3). Strong differences were demonstrated by seed populations collected

Table 5.3 AOSs and non-survival of seeds, frequencies of both cells with chromosomal abnormalities (CCAs) in meristem and chromosomal abnormalities (CAs) in cells

Site	Number of seeds	Number of ana-telophases	AOS ($C_{1/2}$)	Non-survival $1-S$,%	CCA frequency, %	CA frequency,%	Mitotic activity
1998							
P1	167	726	0.22	10.8	2.5 ± 0.5	0.05	9.7 ± 0.9
P2	152	942	0.16	34.2	3.1 ± 0.8	0.03	6.0 ± 0.6
P4	156	518	0.20	19.9	2.5 ± 0.6	0.04	6.3 ± 0.7
P7	149	763	0.50	13.4	1.3 ± 0.3	0.02	10.9 ± 1.0
P8	148	1,047	0.80	31.8	3.2 ± 0.8	0.03	7.8 ± 0.8
P9	167	528	0.76	65.3	4.6 ± 1.4	0.07	6.1 ± 1.0
P10	153	231	0.66	29.4	3.2 ± 0.7	0.05	9.7 ± 3.0
P11	153	342	0.83	55.6	4.4 ± 0.9	0.08	7.3 ± 1.0
P12	148	1,805	0.28	12.3	1.2 ± 0.3	0.01	14.9 ± 0.9
1999							
P2	500	2,228	0.16	72.6***	3.2 ± 0.4**	0.04	17.8 ± 1.2***
P3	500	3,827	0.33	32.6	5.4 ± 0.6	0.06	21.9 ± 1.3
P4	500	1,035	0.22	83.6***	6.8 ± 0.9***	0.09	17.5 ± 1.4***
P5	500	2,209	0.25	67.8	6.3 ± 0.6	0.07	15.0 ± 0.9
P6	500	2,385	0.25	71.6	5.1 ± 0.4	0.07	17.9 ± 1.1
P11	500	2,220	0.50	43.6*	5.5 ± 0.5*	0.07	9.8 ± 0.5**
P12	200	832	0.31	37.5***	5.6 ± 0.8***	0.01	8.0 ± 0.6***

Standard error of the frequency of CAs does not exceed 0.01. Comparing 1998 and 1999 data: *p > 0.1; **p > 0.5; ***p < 0.001

in sites P1, P2, P4–P6 (AOS $\sim 0.16 \div 0.25$) and P7–P11 (AOS $\sim 0.50 \div 0.83$). Populations P1, P2, P4–P6 grow in the humid zone of the Volga River, and the P7-P11 sites are located in the steppe (The Ministry of Atomic Power of the RF et al. 1998a). The site P3 (AOS ~ 0.33) is located close to P2 (AOS ~ 0.16) and at an equal distance from the NPP, but the P2 population grows in the low island part of the town of Balakovo, and the P3 plants on the high bank of the continent (Fig. 5.1). The town of Dubna has a reputation as a wet site, and the AOS value of the seeds collected there is not high. All the above demonstrates the inverse relation of AOS to the humidity of the sites where plants grow. These data can be, most probably, explained by the relation between the rain capacity in the ecosystem and low AOS. This assumption is in agreement with the investigations performed by Zhuravskaya et al. (2000) who have shown that the higher humidity correlates with the lower content and species diversity of the low molecular weight antioxidants, although the latter are usually compensated by an increase in other low molecular weight antioxidants and by the activity of the antioxidant protective ferments.

However, the impact of radiation, chemical factors and heat stress are also likely to be reasons for changes in antioxidant activity in plant cells and tissues (Zhuravskaya et al. 2000; Burlakova et al. 1996). Therefore, the combined effect of humidity and imposed anthropogenic factors may be a cause of the decreased antioxidant protection. AOS determinations have shown that the antioxidant protection of the plantain seeds in five sites placed near the Volga is lower than the average one over all the regions (Table 5.3).

Perhaps, the AOS results reflect the radiosensitivity of these seeds because of a decrease in the diversity of the species' antioxidants, the increase in amount and activity of the radiosensitivity of seeds (Zhuravskaya et al. 2000) and sensitivity to heat stress (Rainwater et al. 1996). The radiosensitivity of plant populations increases with dampness in the ecosystem (Preobrazhenskaya 1971; Zhuravskaya et al. 2000).

Averaged biological values: Table 5.3 shows the averaged biological values: AOS, non-survived seeds, frequencies of chromosomal abnormalities (CAs) in meristem cells, the cells with CAs (CCAs) in meristem, and MA. The value of the non-survived seeds grew high in 1999 in comparison with 1998 (from 20–30 % up to 70–80 %) for the populations located near the NPP (P2–P6). In 1999, the MA elevated in meristem of these populations; it is higher than in any population tested in 1998 ($p < 0.05$, ~threefold), and higher than in "control" P11, P12 plants ($p < 0.05$, ~twofold) (Korogodina et al. 2006). We can speculate on different factors and mechanisms that determined the seeds' viability and variability of cytogenetic values. Let us consider the environmental conditions in detail.

In 1998, one could divide the observed populations into two groups. The first one consists of those that could experience the irradiation impact through atmosphere (P1, P2, P4, P12) (AOS ~0.16–0.28); the second group includes the populations which were located in soil-polluted sites (P7–P11) (AOS ~0.5–0.83). The average values of the non-survived fraction, the frequency of both CCAs and CAs, and MA were not high in populations P1, P2, P4 on the Volga bank (Table 5.3). In the populations located in soil-polluted sites, the $(1-S)$ values and frequencies of abnormal cells and chromosomes can be maximal (P9, P11) (Table 5.3). In 1998, a strong correlation of the $(1-S)$ value with both frequencies of CCAs ($|r_{1-S,\,CCA}| = 0.92$, df $= 7$, $p < 0.001$) (Korogodina et al. 2006) and CAs ($|r_{1-S,\,CA}| = 0.76$, df $= 7$, $p < 0.02$), was observed (Fig. 5.2a). The correlation is observed for populations growing near the Volga bank in the direction of the wind rose (small AOS values) as well as for steppe populations (high AOS values). We can suspect that survival of these seeds was regulated by repair mechanisms in general. It is necessary to mention that strong elevated MA in P12 plants did not disturb this correlation. In the P12 population, a low level of the $(1-S)$ value and both frequencies is coupled with a significant increase of MA, which could be stimulated by a radiation effect (Luchnik 1958) because the neutron dose level was elevated for 2 months that year.

In 1999, the Volga-side populations (P2–P6), which experienced the greatest impact of the NPP fallout,[2] as well as P12 (JINR territory) and P11 (Chernobyl track), were studied. For the populations located near the NPP (P2–P6), the average $(1-S)$ value grew high and the frequency of both CCAs and CAs were elevated in comparison with 1998. These data indicated that plant populations in a 20-km radius of the NPP had experienced high pressure, which decreased viability, increased

---

[2]The tested sites were chosen along the wind rose direction that allows one to study the NPP fallout dose dependence.

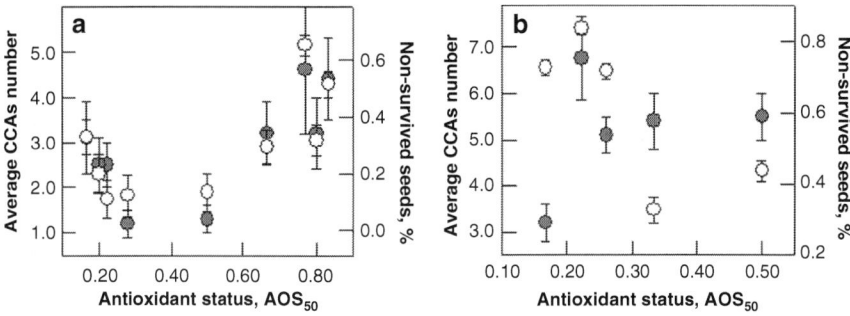

Fig. 5.2 Dependence of the average number of CCAs (*left* Y-axis) and value of non-survival of seeds (*right* Y-axis) on AOS (X-axis) in 1998 (**a**) and 1999 (**b**). Average CCA number and value of non-survival of seeds are shown as *black and white circles*, respectively. The standard errors are shown.

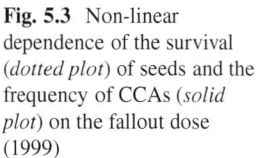

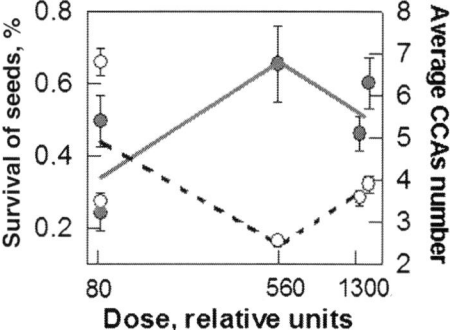

Fig. 5.3 Non-linear dependence of the survival (*dotted plot*) of seeds and the frequency of CCAs (*solid plot*) on the fallout dose (1999)

frequency of abnormalities and stimulated MA. The dependence of these values on the NPP fallout dose has a non-linear character (Fig. 5.3). The data on P11 was the same for both years. This was expected because the population located in chronic radiation conditions had become more resistant in some generations (Shevchenko et al. 1992). For seeds of the P12 population, the $(1 - S)$ value and abnormalities frequency are in the ordinary ratio with reasonable MA elevation.

Let us analyze the correlation between the $(1 - S)$ value and the frequency of CCAs for the 1999 data. In 1999, the correlation between the $(1 - S)$ value and the frequency of CCAs disappeared ($|r| - 0.02$), and the correlation with the frequency of CAs was non-confident statistically ($|r_{1-S, CA}| = 0.51$, df = 5) (Fig. 5.2b). These data indicated that some mechanisms had operated in the seeds collected near the NPP in 1999.

In 1999, the described effects could be induced by both radiation factors and heat because higher temperatures were observed over the whole region of central

5.4 Scheme of Adaptation Processes in Meristem and Cells

Table 5.4 A comparison of the data for the P2 and P3 populations

Parameter	P2	P3
AOS	0.16	0.33
CCA frequency, %	3.2 ± 0.4	5.4 ± 0.6
Non-survival, %	72.6	32.6

Russia.[3] The increasing of $(1 - S)$ as well as of MA values in populations around the NPP differed significantly from those which were observed in distant sites (P11, P12) (Table 5.3), and we can assume the influence of radiation in the 20-km radius of the NPP. For these populations, the effects probably occurred through a combination of atmospheric NPP pollutions (The Ministry of Atomic Power of the RF et al. 1998a) and extreme summer temperatures in 1999 (State Committee on the Environment Protection in the Saratov Region 2000). The hypothesis of the combined effects of radiation and heat stresses was verified in the experiments on pea seeds (Korogodina et al. 2005) (see Sect. 4.3.4).

The systematic radioactive NPP fallout did not exceed the average radiation level over the region (The Ministry of Atomic Power of the RF et al. 1998a; State Committee on the Environment Protection in the Saratov Region 2000) and thus could not have led to mid-lethal DNA damages, but they could exceed it in terms of dose rates and thus induce stress, which causes changes in cells through radiation-induced ROS. Involvement of ROS was verified indirectly by the increase of both the frequency of CAs and $(1 - S)$ depending on the seeds' AOS in populations growing in the town of Balakovo (P2, P3). The lower AOS corresponds to the highest cell elimination and S decreasing. This phenomenon is shown in comparison with the data for the P2 and P3 populations, which grew close to each other (700 m) in the town of Balakovo and are equidistant from the NPP (Table 5.4).

It is mentioned above that there was no correlation between the non-survival and the frequency of CCAs, but the correlation between the AOS and $(1 - S)$ values was found ($|r| \geq 0.75$, $p < 0.05$, $df = 5$). This is the adaptation scheme of survival: to survive, the plant needs to tolerate some mutations, so the adaptation process depends on sensitivity of the plant coupled with the plant's ability to weather the instability processes. Therefore, a high frequency of CCAs and seed survival relate to high AOS value and correspond to the P3 population.

5.4 Scheme of Adaptation Processes in Meristem and Cells

At germination of seeds, the adaptations of rootlet meristems, as well as meristem cells, are required by the environment, which can result in some changes (reconstructions) in meristems and cells (Fig. 5.4). Irradiation of seeds can be completed

---

[3] We don't consider the influence of chemical pollutions because they were decreased in 1999 and we tested populations growing far from their sources (near the Volga). The detailed analysis of possible influence of chemical pollutions is presented in Sect. 8.5.

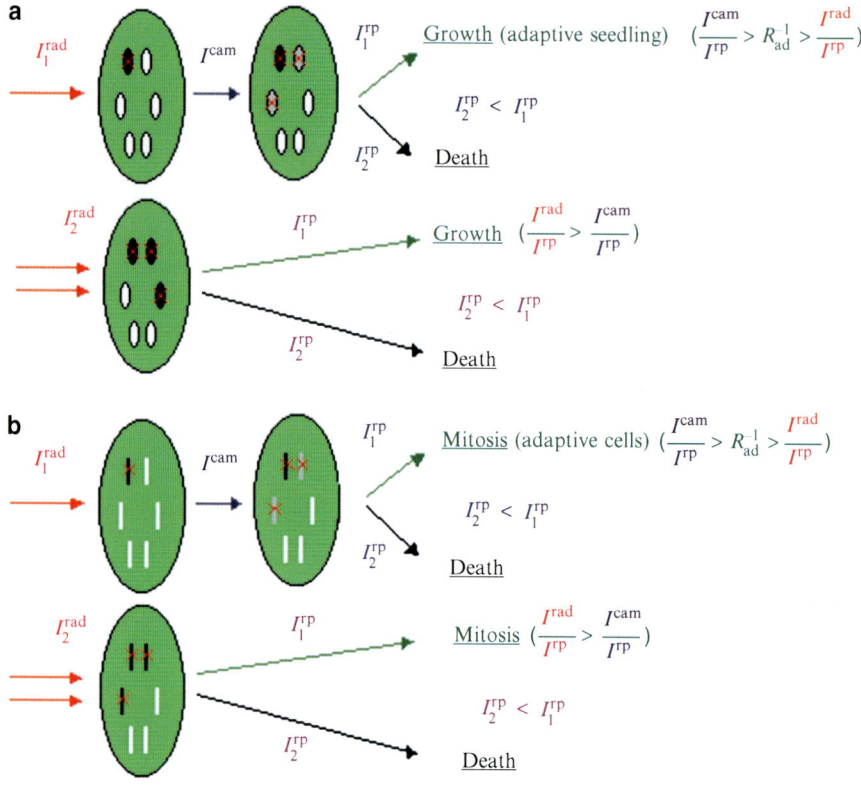

Fig. 5.4 The processes in meristems (**a**) and cells (**b**) (Korogodina and Florko 2007). I^{rad} is intensity of irradiation; I^{com} is intensity of communications; I^{rp} is intensity of repair. Primary damaged cells (**a**) or chromosomes (**b**) are *black*, communicative ones are *grey*, and undamaged ones are *white*

by the stage of first mitoses, considered an "adapted stage of seedling" (Fig. 5.4a), or their death. Some of the irradiated cells come into mitosis, which can be also considered "the adapted stage of cells" (Fig. 5.4b), and the others die.

The reconstructions, which are necessary for adaptation, can be provided by communication processes ($I^{com}/I^{rp2} > R_{ad} > I^{rad}/I^{rp1}$) or primary damages ($I^{rad}/I^{rp1} > I^{com}/I^{rp2}$), where I^{rad} is intensity of irradiation; I^{com} is intensity of communications; I^{rp1}, I^{rp2} are intensities of repair. The abnormalities lead to death of seedling or cell if repair systems are not effective.

In mathematical terms, this process consists of the following components:

- Primary damages of cells or chromosomes which are rare and independent events and can be described by Poisson distribution;
- The communication processes which induce the appearance of secondary abnormalities. This increases Poisson parameter a;

- These reconstructions in meristems and cells provide selection of "adapted" seedlings and cells. A waiting period of adaptation is exponentially distributed resulting in the geometrical distribution of the reconstruction numbers.

It suggests that the appearance of number n abnormalities could include Poisson and geometrical components.

5.5 Statistical Modelling of the Appearance of Cells with Abnormalities, Abnormal Chromosomes in Cells and Proliferated Cells (PCs)

5.5.1 Distributions of Seeds on the Number of Cells with Abnormalities and Cells on the Number of Chromosomes with Abnormalities

The parameters of distributions of seeds on the number of cells with abnormalities and cells on the number of CAs (1998, 1999) are demonstrated in Table 5.5 (see Sect. 8.4.1). All seed distributions have the Poisson and geometric components, with Poisson being predominant in 1998. In 1999, the P-relative value decreased and the sample mean increased. This was expected because the heat stress increased (1999) the appearance of the ROS (Rainwater et al. 1996), which induced the DNA damages (Janssen et al. 1993) and apoptosis (Davies 2000). The combined effect of heat and radiation stress leads to instability processes which decrease the P-subpopulations.

In 1998, the cell distributions on the number of CAs have one geometric component for all the tested plant populations except the two growing in high chemical-polluted sites (State Committee on the Environment Protection in the Saratov Region 2000), which have two geometric components. In the town of Dubna, the plantains are growing near the JINR accelerator facility, and their meristem cells are P-distributed. In 1999, all distributions consisted of two-geometric or geometric plus Poisson components. The Poisson component was observed in the populations growing at the border of the sanitary zone, in the Chernobyl trace territory and near the accelerator.

It is interesting that the cell distributions on the number of CAs can be geometric whereas seed distributions on the number of CCAs could have the Poisson component. We explain this fact by more intensive irradiation of meristem as a whole than of each separate cell. It suggests that irradiation effects are more pronounced in meristems than in cells.

Table 5.5 Parameters of the distributions of plantain seeds on the cells with abnormalities and the distributions of meristem cells on the CAs (Korogodina and Florko 2007)

Seeds	Distribution of seeds on the number of cells with abnormalities				Distribution of cells on the number of CAs					
	^{s}mG**	^{s}G**	^{s}mP*	P*	^{c}mG1*	^{c}G1*	^{c}mG2*	^{c}G2*	^{c}mP*	^{c}P*
1998										
P1	1.10	0.17	0.22	0.41	0.01	0.42	0.30	0.07		
P2	0.05	0.10	0.17	0.45	0.03	0.26				
P4	0.07	0.06	0.17	0.64	0.04	0.38				
P7	0.14	0.27	0.19	0.38	0.02	0.52				
P8	0.10	0.07	0.26	0.39	0.03	0.34				
P9	2.40	0.01	0.36	0.19	0.07	0.39	0.28	0.04		
P10	2.30	0.04	0.32	0.46	0.03	0.35				
P11	0.70	0.09	0.45	0.21	Non-confidence statistically data					
P12	0.06	0.12	0.21	0.69					0.01	0.9
av.	0.77	0.10	0.26	0.42						
1999										
P2	1.05	0.13	0.36	0.11	0.02	0.43	0.14	0.12		
P3	1.85	0.30	1.20	0.37	0.02	0.40	0.11	0.55		
P4	2.01	0.02	1.26	0.12	0.03	0.25			0.53	0.03
P5	0.08	0.04	1.20	0.26	0.06	0.50	0.23	0.05		
P6	2.45	0.04	0.96	0.21	0.02	0.36			0.22	0.12
P11	0.32	0.10	0.84	0.33	0.02	0.36			0.20	0.17
P12	1.81	0.12	0.36	0.39	0.02	0.19			0.35	0.03
av.	1.36	0.11	0.89	0.26						

*Standard errors of the parameters do not exceed 20–30 % (the sample means) and 10–15 % (the relative values); **Standard errors of the parameters do not exceed 40–50 % (the sample means) and 20–30 % (the relative values)

5.5.2 Distribution of Seeds on the Number of PCs in Seedling Meristem

Let us consider stimulation of proliferation of resting cells, which significantly contributes to adaptation. Stimulation of proliferation of resting cells is not a compensatory mechanism (Luchnik 1958) and the number of PCs is an independent value. We assume that the number of cells in a stationary phase can be described as a stationary random branch process (see Sect. 3.4.3). It follows from the above that the MA of cells in meristem can be described by the lognormal (LN) distribution of cells on the number of PCs (Florko and Korogodina 2007)[4] in the first mitoses phase. This conclusion is in agreement with the LN law described by A.N. Kolmogorov (1986). In the case of intensive selection, the LN distribution can be transformed into a geometric distribution (Florko and Korogodina 2007).

---

[4]The proof has been obtained together with V.B. Priezzhev

5.5 Statistical Modelling of the Appearance of Cells with Abnormalities...

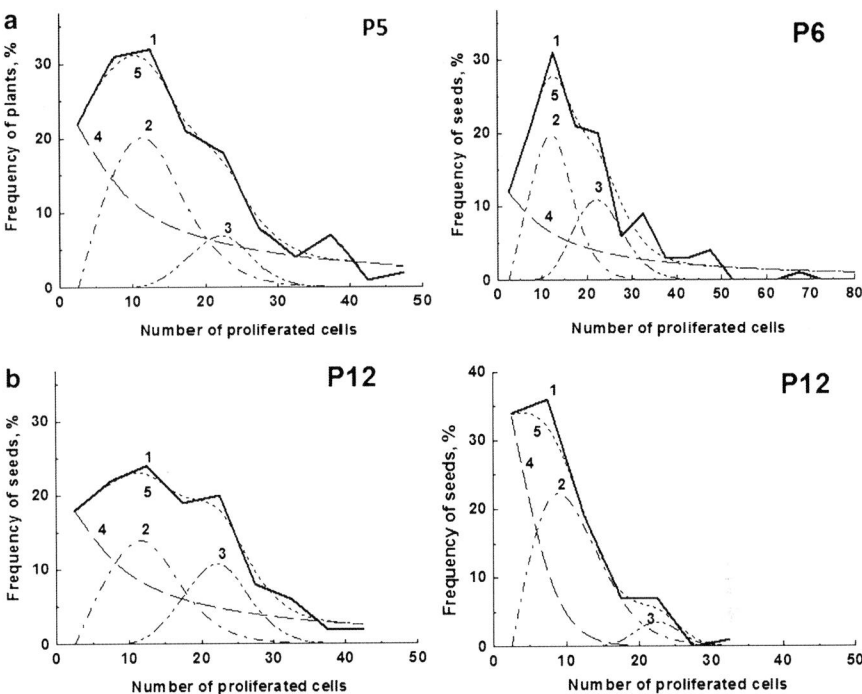

Fig. 5.5 (**a**) The approximations of the experimental distributions of plantain seeds on the number of PCs. Seeds were collected in the P5 and P6 populations (1999) *Plots: 1*, experimental data; *2, 3, 4*, geometric or lognormal modelling components; *5*, sum of the modelling plots. (**b**) The approximations of the experimental distributions of plantain seeds on the number of PCs. Seeds were collected in the P12 population in 1998 (*left*) and 1999 (*right*) (**b**). *Plots: 1*, experimental data; *2, 3, 4*, geometric or lognormal modelling components; *5*, sum of the modelling plots

Approximations of the experimental distributions of seedlings on the number of PCs were performed by using criteria χ^2, AIC, BIC, T. Their analysis has shown a sum of three LN distributions 3LN or a sum of two – the LN plus the geometric ones 2LN + G (see Sect. 8.4.2). The geometric component is revealed in the cases of the P5, P6, and P7 populations growing near the NPP (Fig. 5.5a) and P12, which grows in the JINR territory (Fig. 5.5b). The relative value and sample mean of the geometric component are 45–47 % and 33, respectively, for P5 (1999) and P12 (1998) plant populations in comparison with the same for P12 in 1999 – 30 % and 4.

5.5.3 Regularities of the Adaptation-Processes Induced by Nuclear Station Fallout in Plantain Populations

The parameters of the statistical model (1999) are presented in Figs. 5.6, 5.7, and 5.8. The parameters of seed distributions (Fig. 5.6) and cell distributions (Fig. 5.7)

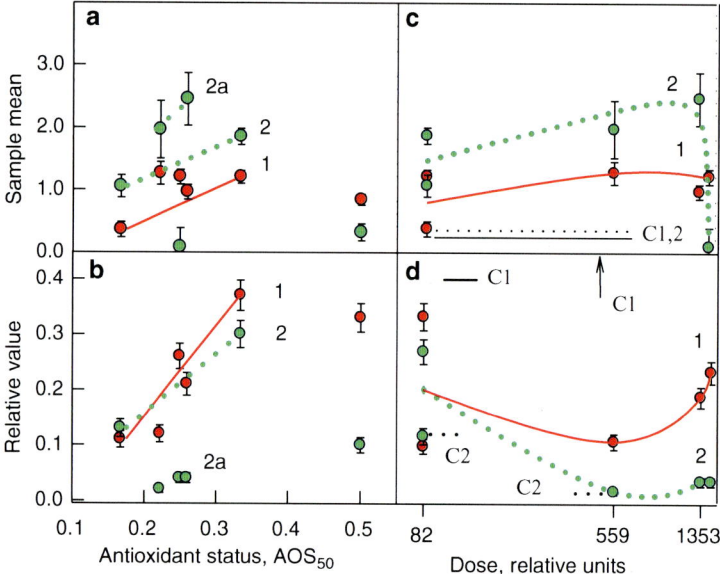

Fig. 5.6 The parameters of ^sP and ^sG distributions of plantain seeds on the number of CCAs in root meristem of seedlings (Korogodina et al. 2006) versus AOS (**a, b**) and the calculated relative dose (**c, d**). The sample mean (**a, c**) and value (**b, d**) of the ^sP and ^sG distributions are shown (pl. 1 and 2, respectively). Points with the same dose-rate irradiation are connected (**a, b**). ^sP and ^sG (1998) are designed by the *solid* and *dotted lines* (**c1,2**), respectively (**c, d**). The regressions, **a**: y = −0.48 + 5.04x (p < 0.05) (*1*), y = 0.25 + 4.80x (p < 0.05) (*2*); y = −0.92 + 12.96x (p < 0.05) (*2a*); **b**: y = −0.15 + 1.43x (p < 0.05) (*1*), y = −0.04 + 1.02x (p < 0.05) (*2*); y = −0.10 + 0.53x (p < 0.05) (*2a*); **c**: polynomial fits (*1, 2*); **d**: polynomial fit (*1*); y = 100/x$^{1.4}$ + 0.027 (*2*)

are shown as dependent of the seeds' AOS (Figs. 5.6a, b and 5.7a, b) and the calculated fallout dose near the NPP (Figs. 5.6c, d and 5.7c, d) (Korogodina et al. 2010a, b).

The parameters of the $^cG2$ and ^cP distributions characterize the damage for cells of the sensitive subpopulation (Table 5.5) and therefore they are combined (Fig. 5.7). These parameters are correlated with ^sG value of the sensitive subpopulation (p < 0.001, df = 5, n = 7), which consists of more resistant survived seeds of the sensitive subpopulation (Korogodina et al. 2010a). Thus, the ^sG subpopulation can be considered as the instability-risk group which experiences selection.

The values of sensitive subpopulations of surviving seedlings (Fig. 5.6d; pl. 2) and cells (Fig. 5.7d; pl. 2) decrease with the fallout dose rate (p < 0.1, df = 3, n = 5). At the border (558 and 1,350) or inside (1,500 r.u.[5]) the 10-km zone, approximately

---

[5]Evaluation of this relative dose rate can be less reliable because this population was located ≈ 100 m from the NPP in the shade of the smokestack.

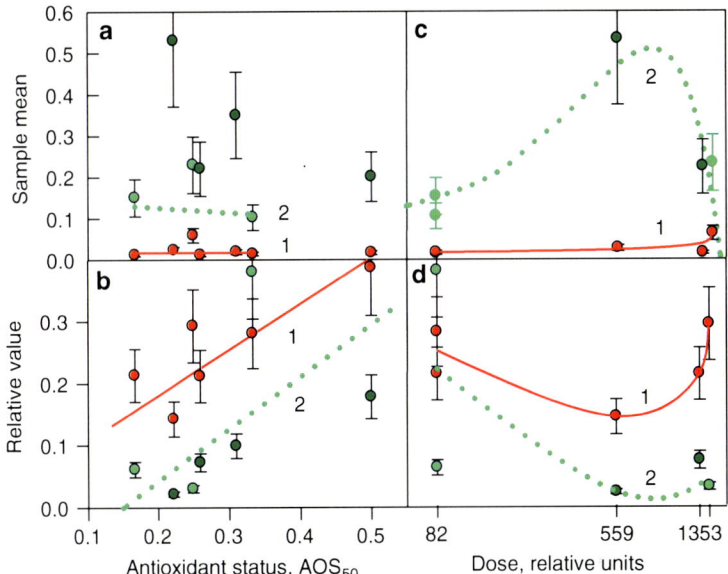

Fig. 5.7 The parameters of $^cG1$ and $^cG2$, ^cP distributions of plantain meristem cells on the number of CAs (Korogodina and Florko 2007) for the seeds collected around the NPP versus AOS (**a**, **b**) and the calculated relative dose (**c**, **d**). The sample mean (**a**, **c**) and value (**b**, **d**) of the $^cG1$, $^cG2$ and ^cP distributions are shown (pl. 1 and 2, respectively). Points with the same dose-rate irradiation are connected (**a**, *1*, *2*). The regressions, **b**: $y = 0.73x + 0.02$ ($p < 0.05$) (*1*), $y = 0.77x - 0.12$ ($p < 0.05$) (*2*); **c**: polynomial fits (*1*, *2*); **d**: polynomial fit (*1*); $y = 100/x^{1.4} + 0.027$ (*2*)

70–80 % of seeds died, and the value of the selected seedlings ^sG is about 2–3 %. Meristem cells of these survived adapted seedlings consist of the increased number of late and primary DNA damages ($p < 0.05$) (Figs. 5.6c, d and 5.7c, d; pl. 2).

The data related to the distributions of seeds collected in 1998 were used as the control (Fig. 5.6c, d, lines C1, 2). One can see that both sample means of ^sP and ^sG subpopulations increased in 1999 in comparison with 1998 (Fig. 5.6c), which we can explain by the increased instability. The ^sP value decreased in 1999, whereas the ^sG one was preserved (Fig. 5.6d). It can be assumed that increased instability decreases the Poisson subpopulation, but the geometric subpopulation does not decrease because resistance of seeds is constant. Finally, the fraction of non-survived seeds increased (Table 5.3). We can suspect the combined effect induced by heat and radiation stresses in 1999.

The comparison of the frequency of CCAs and non-survival of seeds growing in the P2 (low AOS value) and P3 (high AOS value) plantain populations presents us with a paradox: the frequency of CCAs is lower but non-survival is higher for the P2 population where AOS value is lower (Table 5.4). It can be added now by the calculated modelling parameters (Table 5.6). The sample mean is higher in the P3 resistant population than in the P-law and the geometric one. It follows from the above that survival is higher in the resistant population and relative values should

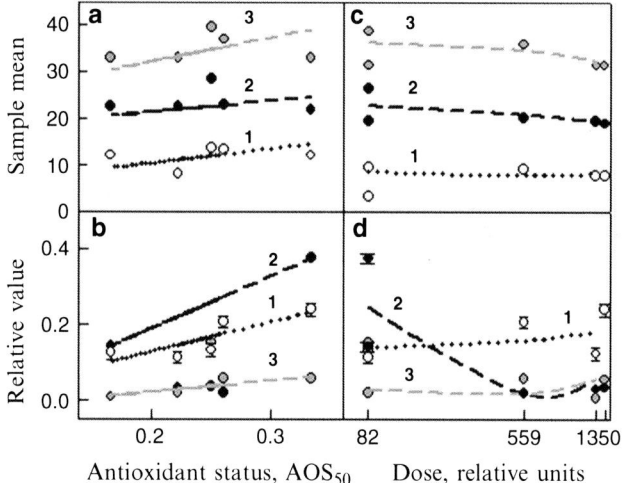

Fig. 5.8 Dependence of sample mean (**a, c**) and relative value (**b, d**) of seeds' distributions on the numbers of PCs in meristem of rootlet on AOS (**a, b**) and relative dose (**c, d**). Plots of parameters of three LN distributions are marked with *1*, *2*, and *3*

Table 5.6 A comparison of the seeds distribution parameters for the P2 and P3 populations

Parameter	P2	P3
AOS	0.16	0.33
Non-survival	72.6	32.6
^sP	0.11 ± 0.01	0.37 ± 0.04
^smP	0.36 ± 0.10	1.20 ± 0.02
^sG	0.13 ± 0.01	0.30 ± 0.02
^smG	1.05 ± 0.17	1.85 ± 0.13

also be greater (Table 5.5). It is demonstrated in Fig. 5.6 as well. For the seeds collected in the P2, P3 populations of neighboring ecosystems (relative NPP fallout dose rate = 82 r.u.), the parameters of the ^sP and ^sG distributions (Fig. 5.6b, pl. 1, 2) and the $^cG2$ distribution (Fig. 5.7b, pl. 2) differed ($p < 0.05$) in dependence on the seeds' AOS. The picture shows increasing of the sample mean of ^sG-subpopulation with the AOS ($p < 0.05$) (Fig. 5.6a, pl. 2).

Dependence of appearance of PCs on dose rate irradiation: The PCs can appear in the P- and G-subpopulations due to heterogeneity of the proliferative pool, and the third subpopulation corresponding to the activated resting cells (Florko and Korogodina 2007). The analysis of the distribution of seedlings on the number of PCs has shown preference for the 2LN + G model on χ^2, AIC, BIC, and T criteria (see Sect. 8.4.2). The 3LN model meets the χ^2-criterion, and their parameters can be used to investigate the dependence of the appearance of PCs on the NPP fallout and AOS (Fig. 5.8).

Parameters of all distributions depend on the AOS (Fig. 5.8a, b), but the second subpopulation is the most sensitive (Fig. 5.8b). Dependence on the relative dose rate is presented in Fig. 5.8c, d. The values of LN1 and ^{s}P, as well as LN2 and ^{s}G are correlated: $|r| = 0.83$ (df $= 5$, $p < 0,01$) and $|r| = 0.95$ (df $= 4$, $p < 0.01$), respectively (Florko and Korogodina 2007). It suggests that distributions LN1 and LN2 describe resistant and sensitive subpopulations of the PCs. The LN3 contributes significantly in the MA, which is related to both ^{s}G- and LN3- distributions (homogeneity criterion, $\chi^2 = 2.9$; df $= 4$; $p << 0.001$). The value of the LN3-distribution increases significantly ($p < 0.05$) in the sanitary zone of the NPP (Fig. 5.8d), thereby it elevates both the number of PCs and CAs (because most of the resting cells are mutant (Bridges 1997)).

5.6 Two Evolution Strategies of Survival

In 1998, we could roughly separate two groups of populations: those at the riverside and those at the steppe (Fig. 5.2a). The populations are capable of being attacked by radiation in the first case and by chemical agents in the second. A strong correlation of the $(1 - S)$ value with the frequency of both CCAs and CAs was mentioned above (Sect. 5.3). The general picture is that the low AOS value relates to the high number of abnormalities and to decreased survival of seeds (Fig. 5.2a). That means inverse relation of the number of CAs and survival, when the main instrument of survival is cellular repair. This is in agreement with the absence of late processes in cells because ^{c}G2- and ^{c}P-components of cells' distributions were not revealed practically for all populations (Table 5.5).

In 1999, such relations disappeared (Sect. 5.3, Fig 5.2b). The strong late processes of chromosomal instability were observed which were revealed as ^{c}G2- and ^{c}P-components of cell distributions (Table 5.5). The MA value was also increased (Table 5.3) due to stimulation of resting cells (Fig. 5.5) that contributed to survival and material for selection too. To conclude, we see the adaptation processes induced by radiation-heat stresses: the chromosomal instability accompanied by selection.

This example demonstrates methods to survive: the first strategy uses the chromosomes' reparation which was preferable in 1998, and the second one is based on the adaptation to stress (1999). In both cases, the numbers of abnormalities increase whereas the old genotype eliminates, especially while looking for new adaptive genotypes. In 1939, N.W. Timofeeff-Ressovsky pointed out for the first time (Timofeeff-Ressovsky 1939) that microevolution process includes both a necessary increase of the material for evolution and decrease of a part of the old population. Then, M. Eigen and P. Schuster (1979) showed that elimination of organisms with old genotypes is necessary for survival of the new ones in the view of living resources of the adapted population.

5.7 Risk of Chromosomal Instability

It is of interest to calculate the risk of the chromosomal instability which always accompanies genetic adaptation induced by the radiation stress factor. It can be estimated by means of statistical modelling. The fitting divides the seeds' population into two subpopulations. The Poisson subpopulation corresponds to the primary and late cells' injuring, and the geometric one means the appearance of the same accompanied by selection.

At the border of the sanitary zone, the number of abnormal cells was negligible in 1998 (Fig. 5.6c, pl. C 1, 2). The higher temperature (1999) increased it up to approximately one cell with abnormalities per meristem in fifty percent of seedlings for the P subpopulation and more than two cells for the G-one (Fig. 5.6c). It is enough to decrease the seed number from 80 to 20 % in the P subpopulation (Fig. 5.6d). In the resistant fraction, the risk of injuring of cells can be calculated with P parameter as 20 %. In the sensitive fraction, the risk of instability is accompanied by selection and can be calculated with G-parameters as 1–2 %. The risk of seed death is approximately 80 %. Risks of the bystander and chromosomal instability processes increase with the decline of AOS (Fig. 5.6c), and increase with the fallout dose (Fig. 5.6d).

5.8 Summary

Effects of low-dose radiation comparable with the background are always debated. It is difficult to register such effects, especially in nature, due to the influence of other factors. These ecological investigations can be performed near the NPP because their fallout does not exceed the background. The Balakovo NPP was chosen as a radiation source because blood diseases in Balakovo inhabitants are well-documented (Dodina 1998). The plantain seeds were used as the investigation object since we can find natural populations having experienced different radiation effects. The sources of the chemical pollutions are located in the Balakovo region which should be taken into consideration. The method of statistical modelling is based on the experimental data and has proved effective in these problems. The statistical modelling divides the population into resistant and sensitive subpopulations and model parameters are more sensitive than the averaged biological values.

The influence of the main environmental factors was studied before the analysis of the cytogenetic and survival data. Annual γ-radiation dose rates are $\sim$0.10–0.15 μSv/h that does not exceed the background (The Ministry of Atomic Power of the RF et al. 1998a). It is shown that chemical pollution could influence the plant populations in 1998 as well as high summer temperatures in 1999 (Korogodina et al. 2006).

5.8 Summary

Analysis of the averaged values: The AOS of seeds was studied to characterize the cellular protection mechanisms induced by ROS. It is shown in (Zhuravskaya et al. 2000) that AOS value of plants is determined by humidity of their population site. AOS studies have demonstrated resistance of seeds from the steppe sites and their sensitivity from the populations growing near the Volga River. Two population sites are located at a short distance in the town of Balakovo, but they differ significantly in humidity and therefore on seeds' AOS. Cytogenetic values and survival are different for the seeds collected there: both frequency of abnormal meristem cells in seedlings and their survival are less in the sensitive seeds.

In 1998, the populations growing in the steppe were under chemical influence, and seeds collected near the Volga River experienced radiation impact through atmosphere. In 1999, the tested populations were growing at the Volga River and experienced a combined effect of NPP fallout and high summer temperature. A significant correlation is observed between the seeds' survival and the number of cells with abnormalities in seedlings' meristem in 1998, and this correlation was disturbed in 1999. It supposes different mechanisms which regulated the seeds' survival and the number of cells with abnormalities in these years. We can assume direct dependence of seeds (and cell) death on the number of abnormalities in 1998, and prevalence of the adaptation processes in 1999.

These adaptation processes correct the cell genome by accumulation of abnormalities until features of seedling reach the balance with environmental conditions. The process is determined by intensities of the primary and late damage processes, and plant sensitivity.

Statistical modelling: The modelling was performed to study appearance of both cells with abnormalities and the normal proliferated ones in meristem, and also CAs in meristem cells (see Sects. 8.4.1 and 8.4.2). The modelling parameters correlate among themselves and biological values and indicate the NPP fallout as the factor which induces the instability and selection processes. It is necessary to stress that the LN distribution of seedlings on the number of proliferated root meristem cells transforms into the geometric one near the irradiation sources such as NPPs and the JINR facilities.

Risk of chromosomal instability: As we see, the radiation stress induces chromosomal instability accompanied by selection. The probability of the chromosomal instability and selection can be estimated by Poisson- and geometric values. Risks increase with additional synergic factors such as high temperature. The described processes lead to the changes of the previous genotype on the adaptive one. Man-raised plants and breeds will change first of all because their genotype does not accommodate to natural conditions (Glazko 2006). The second conclusion is decreased numbers of some species right up to their disappearance; such examples are presented in Sect. 3.3. In this case, the species' abundance distribution can be described by the geometric law.

References

Alenitskaja SI, Bulah OE, Buchnev VN et al (2004) Results of long-term supervision over the environment radioactivity in area of JINR arrangement. PEPAN Lett 1:88–96

Amiro BD (1992) Radiological dose conversion factors for genetic non-human biota. Used for screening potential ecological impacts. J Environ Radioactivity 35:37–51

Artyukhov VG, Kalaev VN, Savko AD (2004) Influence of irradiation of parent oak trees (*Quercus robur L.*) on cytogenetic values of seed posterity (delayed consequences). Bulletin Voronezh Univ (Voronezh) 1:121–128 (Russian)

Averbeck D (2010) Non-targeted effects as a paradigm breaking evidence. Mutat Res 687:7–12

Bamblevskij VP (1979) Efficiency of gamma-quanta detection of three-dimension cylindrical sources by NaI(Tl) crystals. Sov J Pribory i Tehnika Experimenta 2:68–72 (Russian)

Bridges BA (1997) DNA turnover and mutation in resting cells. Bioessays 19(4):347–352

Burlakova EB, Dodina L, Zyuzikov N et al (1998) On the problem of low dose ionizing radiation and chemical pollution effects on man and biota. Project "Evaluation of the combined effect of the radionuclide and chemical pollutions". Atomnaya energiya 85:457–462 (Russian)

Burlakova EB, Goloschapov NV, Gorbunova NV et al (1996) In: Burlakova EB (ed) Consequences of the Chernobyl accident: human health. Center of the Ecology Politics of Russia, Moscow 1, pp. 149–182 (Russian)

Davies KJ (2000) Oxidative stress, antioxidant defenses, and damage removal, repair, and replacement systems. IUBMB Life 50:279–289

Dodina LG (1998) The health violation of individuals and adaptation mechanisms in the conditions of anthropogenic low-intensity factors. Dr. Sci. theses. Saint Petersburg State Medical Academy named after II Mechnikov, St-Petersburg (Russian)

Eigen M, Schuster P (1979) The hypercycle. Springer, Berlin/Heidelberg/NY

European Commission (1998) The Atlas of caesium-137 contamination in Europe after the Chernobyl accident. In: Kresson E (ed) DG XII, Safety Programme for Nuclear Fission (Protection against Radiation), Joint Research Centre, Luxemburg

Florko BV, Korogodina VL (2007) Analysis of the distribution structure as exemplified by one cytogenetic problem. PEPAN Lett 4:331–338

Foyer CH, Noctor G (2009) Redox regulation in photosynthetic organisms: signaling, acclimation, and practical implications. Antioxid Redox Signal 11:861–905

Geras'kin SA, Fesenko SV, Alexakhin RM (2008) Effects of non-human species irradiation after the Chernobyl NPP accident. Environ Int 34:880–997

Glazko VI (2006) 20 years after Chernobyl: genetic consequences of the accident. Priroda 5:48–53 (Russian)

Grodzinsky DM (2009) Chernobyl catastrophe. Foreword. Ann N Y Acad Sci 118:vii–ix

Grodzinsky DM, Kolomiets KD, Kutlakhmedov YA et al (1991) Anthropogenic radionuclide anomaly and plant. Lybid, Kiev

Grodzinsky DM, Kravets EA, Khvedynich OA et al (1996) Formation of the reproduction plant system. Tsytologiya i genetika 30:37–45 (Russian)

Grodzinsky DM (1999) General situation of the radiological consequences of the Chernobyl accident in Ukraine. In: Imanaka T (ed) Recent research activities about the Chernobyl NPP accident in Belarus, Ukraine and Russia. University Research Reactor Institute (KURRI-KR-7), Kyoto, pp 18–28

Gusev NG, Beljaev VA (1986) Radionuclide releases into biosphere. EnergoAtomIzdat, Moscow (Russian)

Hosker RP (1974) Estimates of dry deposition and plume depletion over forests and grasslands. In: Proceedings of physical behavior of radioactive contaminants in the atmosphere. IAEA, Vienna 291–309

Ivanov VI, Lyszov VN, Gubin AT (1986) Handbook on microdosimetry. Atomizdat, Moscow (Russian)

Janssen YM, Van Houten B, Born PJ et al (1993) Cell and tissue responses to oxidative damage. Lab Invest 69:261–274

Kolmogorov AN (1986) About the log-normal distribution of particle sizes under fragmentation. In: The probabilities theory and mathematical statistics. Nauka, Moscow (Russian)

Korogodina VL, Bamblevsky V, Grishina I et al (2000) Antioxidant status of seeds collected in the plantain *Plantago major* populations growing near the Balakovo nuclear power plant and chemical enterprises. Radiats Biol Radioecol 40:334–338 (Russian)

Korogodina VL, Bamblevsky V, Grishina I et al (2004) Evaluation of the consequences of stress factors on plant seeds growing in a 30-km zone of Balakovo NPP. Radiats Biol Radioecol 44:83–90 (Russian)

Korogodina VL, Florko BV, Korogodin VI (2005) Variability of seed plant populations under oxidizing radiation and heat stresses in laboratory experiments. IEEE Trans Nucl Sci 52:1076–1083

Korogodina VL, Bamblevsky CP, Florko BV et al (2006) Variability and viability of seed plant populations around the nuclear power plant. In: Cigna AA, Durante M (eds) Impact of radiation risk estimates in normal and emergency situations. Springer, Dordrecht, pp 271–282

Korogodina VL, Florko BV (2007) Evolution processes in populations of plantain, growing around the radiation sources: changes in plant genotypes resulting from bystander effects and chromosomal instability. In: Mothersill C, Seymour C, Mosse IB (eds) A challenge for the future. Springer, Dordrecht, pp 155–170

Korogodina VL, Florko BV, Osipova LP (2010a) Adaptation and radiation-induced chromosomal instability studied by statistical modeling. Open Evol J 4:12–22

Korogodina VL, Florko BV, Osipova LP et al (2010b) The adaptation processes and risks of chromosomal instability in populations. Biosphere 2:178–185 (Russian)

Kovalchuk O, Dubrova YE, Arkhipov A et al (2000) Wheat mutation rate after Chernobyl. Nature 407:583–584

Lozovskaya EL, Sapezhinskii II (1993) Comparative efficiency of some medicinal preparations as superoxide acceptors. Biophysics 38:25–29 (Russian)

Luchnik NV (1958) Influence of low dose irradiation on mitosis in pea. Newsletter of Ural Department of Moscow Society of Nature Investigators 1:37–49 (Russian)

Maslov OD, Belov AG, Starodub GY et al (1995) Activation analysis of environmental samples using the MT-25 Microtron of the FLNR. In: Abstracts of the third Asian conference on analytical sciences, Seoul, 1995, p. 217

Preobrazhenskaya E (1971) Radioresistance of plant seeds. Atomizdat, Moscow (Russian)

Rainwater DT, Gossett DR, Millhollon EP et al (1996) The relationship between yield and the antioxidant defense system in tomatoes grown under heat stress. Free Radic Res 25:421–435

Romanovskaya VA, Sokolov IG, Rokitko PV et al (1998) Ecological consequences of radiation pollution for soil bacteria in 10-km zone of the Chernobyl Atomic Station. Mikrobiologiya 67:274–280

Shevchenko VA, Pechkurenkov VL, Abramov VI (1992) Radiation genetics of the native populations. Genetic consequences of the Kyshtym accident. Nauka, Moscow

Shevchenko VV, Grinikh LI, Shevchenko VA (1995) The cytogenetic effects in natural populations of *Crepis tectorum* exposed to chronic irradiation in the region of the Chernobyl Atomic Electric Power Station. An analysis of the frequency of chromosome aberrations and karyotype changes in the 3rd and 4th years after the accident. Radiats Biol Radioecol 35(5):695–701 (Russian)

Shevchenko VA, Abramov VI, Kal'chenko VA et al (1996) The genetic sequelae for plant populations of radioactive environmental pollution in connection with the Chernobyl accident. Radiats Biol Radioecol 36(4):531–545 (Russian)

Shevchenko VA (1997) Integral estimation of genetic effects of ionizing radiation. Radiats Biol Radioecol 37(4):569–576 (Russian)

State Committee on the Environment Protection in the Saratov Region (2000) Conditions of environment in the Saratov region in 1999. Acvarius, Saratov (Russian)

The Ministry of Atomic Power of the RF, Rosenergoatom concern, Balakovo NPP (1998a) The general information on Balakovo NPP. Balakovo NPP, Balakovo (Russian)

The Ministry of Atomic Power of RF, Rosenergoatom concern, Balakovo NPP (1998b) The basic technical characteristics of Balakovo NPP energy blocks. Balakovo NPP, Balakovo (Russian)

Timofeeff-Ressovsky NW (1939) Genetik und Evolution. Zeitschrift für inductive Abstammungs and Vererbungslehre 76:158–218 (Russian)

Wiegel B, Alevra AV, Matzke M et al (2002) Spectrometry using the PTB neutron multisphere spectrometer (NEMUS) at flight altitudes and at ground level. Nucl Instrum Methods A 476:52–57

Yablokov AV, Nesterenko VB, Nesterenko AV (2009) Chernobyl: consequences of the catastrophe for people and the environment. Ann N Y Acad Sci 1181:vii–xiii, 1–327

Zhuravskaya A, Stognij V, Kershengolts B (2000) Influence of the environment conditions on the seeds antioxidant system activity of different plant species. Rastitelnye resursy 36(1):57–64 (Russian)

Zykova AS, Voronina TF, Pakulo AG (1995) Radiation situation in the Moscow and Moscow region caused by fallouts in 1989–1993. Gigiena i Sanitarija 2:25–27 (Russian)

Chapter 6
Instability Process Across Generations. Consequences of Nuclear Test Fallout for Inhabitants

Abstract The radiation effects on the human populations living in regions distant from the sites of nuclear explosions that took place in the middle of the previous century are analyzed. The statistical modelling was performed to study the occurrence frequency of abnormal lymphocyte cells among the proliferated ones in the blood of individuals living in the Yamalo-Nenets autonomous district (North Siberia), and settlements in Maloe Goloustnoe and Listvyanka (Pribaikal'e). Four generations of individuals were tested. It was shown that the geometric model component corresponds to the individuals with bad activated lymphocyte cells, lymphocyte pool depletion, and increased mortality, and the Poisson model means accumulation of abnormal cells. The Poisson component was only revealed in younger generations and can be interpreted as "effect of youth." The worst situation is observed in the Northern population, which can be expected due to Northern permafrost and the traditional food chain of "lichen-reindeer-man". The influence of the radiochemical industry on the occurrence of multi-aberrant cells in the blood of its workers and the inhabitants of the town in which it operates was studied by the statistical modelling, with elevated chromosomal instability being found. We conclude that chromosomal instability induced by nuclear test fallout continued for four generations. It was shown that the Poisson sample mean decreased very slowly across a generation which disputes the opinion that reduction of cellular instability in youngsters in the previous investigations was based on the averaged values. In addition, aging and extreme conditions increase the risks of chromosomal instability and mortality.

Keywords Nuclear test fallout • Northern extreme conditions • Statistical modelling • Aging • Chromosomal abnormalities • Transgenerational instability • Risks of chromosomal instability • Radiochemical industry

6.1 General Genetic Consequences of Radiation Impacts

The main consequences of dramatic radiation catastrophes have been studied by examples of the Chernobyl accident and the Semipalatinsk nuclear explosions. It is important to compare the late consequences for human populations living in the radiation-polluted sites located close to the radiation source to those living at a great distance away from it.

The examples will be considered as to how the nuclear tests performed in the Novaya Zemlya and Semipalatinsk polygons in the middle of the previous century influenced the Siberian inhabitants in the Polar Tundra and the Pribaikal'e region. Perhaps the statistical modelling allows one to analyze some additional features of the instability process in this particular generation because it "divides" the whole population into resistant and sensitive fractions.

6.1.1 Consequences of Dramatic Nuclear Impacts

Scientists have performed a considerable number of investigations into the chromosome analysis of blood lymphocytes of individuals living in the radiation-polluted sites or those which were directly irradiated due to nuclear tests in the Semipalatinsk polygon, and by the accidents in the South Urals and Chernobyl (Dubrova 2003; Yablokov et al. 2009; Shevchenko and Snigiryova 1999). V.A. Shevchenko (1997) demonstrated the same genetic regularities on the territories polluted due to the accidents in the nuclear power plants in Chernobyl and Three Mile Island, and those related to the activity of the chemical plant Mayak in the Chelyabinsk region and nuclear tests in the Semipalatinsk site. It is known that irradiation results in genetic damages which induce mutations of different types: genome mutations, chromosome mutations, and small point mutations. Investigations have shown a significant increase of all types of mutations which include the total number of chromosomal aberrations and specific markers of the radiation effect – the rings and dicentrics in human populations living around the Chernobyl area and in the other sites mentioned above in comparison with the control group (Yablokov et al. 2009). The data are published on the genetic radiation-induced congenital malformation and health of children of irradiated parents (Yablokov et al. 2009).

Chernobyl accident: Two years after the Chernobyl accident, I.M. Eliseeva et al. (1994) carried out a cytogenetic study of the children, dwelling in two regions of Ukraine where radiation fallout had occurred. Chromosome analyses of these individuals have shown a significant increase of the frequency of aberrant cells and chromosomal type aberrations in comparison with the control group. The authors have revealed a significant increase of the chromosomal-type aberrations level, extension in the spectrum of complex aberrations of chromosomes (dicentrics, rings and exchange aberrations) and a share of children with various chromosomal abnormalities (CAs) through the years.

6.1 General Genetic Consequences of Radiation Impacts

Table 6.1 Comparative frequency (%, M ± m) of aberrant cells and chromosomal aberrations (per 100 lymphocytes) in different periods (Bochkov et al. 1972, 2001; Pilinskaya et al. 1991; Bezdrobna et al. 2002; Yablokov et al. 2009)

Date	Region	Aberrant cells	Chromosomal aberrations
Before 1986	Ukraine	1.43 ± 0.16	1.47 ± 0.19
	All over the world	2.13 ± 0.08	2.21 ± 0.14
1998–1999	Ukraine	3.20 ± 0.84	3.51 ± 0.97
	30-km zone of the Chernobyl nuclear power plant	5.02 ± 1.95	5.32 ± 2.10

The studies which had been performed by V.A. Shevchenko and G.P. Snigiryova (1999) showed that during the whole period of investigations (1986–2004) the frequency of dicentrics in peripheral blood lymphocytes was significantly higher than the control level. The cytogenetic investigations (1992–1994) of the Bryansk region inhabitants showed its value increased by five times in comparison with the control group.

Over many years, V.A. Shevchenko (Moscow N.I. Vavilov Institute of General Genetics) and his colleagues studied the consequences of the Chernobyl accident in some sites, and showed a high frequency of CAs including the radiation markers which continued across generations (Shevchenko 1997).

In the United Kingdom, Yu.Dubrova et al. (1996) showed that DNA mutation level was on average two times higher in the children who were born in Belarus in 1994 and whose parents lived close to the polluted territories after the Chernobyl accident than in the children in control families in the United Kingdom, and that it was correlated with the radiation pollution level of the territory where their parents had lived. The number of mutations increases not only in somatic but also in generative cells. The level of small mutations in mini-satellite DNA in the children who were born in Belarus and Ukraine and whose parents had lived in the radiation-polluted territories was approximately two times higher than in the children in the United Kingdom (Dubrova et al. 1996; Dubrova 2003).

The monography "Chernobyl: the consequences for man and nature" (Yablokov et al. 2009) presents the data 20 years after the Chernobyl accident. The cited data have demonstrated a significant increase of the frequency of chromosomal aberrations in the polluted areas (Lazjuk et al. 1999; Sevan'kaev et al. 1995; Vorobtsova et al. 1995; Mikhalevich 1999). The comparison of the data on the frequency of aberrant cells and chromosomal aberrations investigated in different periods is presented in Table 6.1.

Semipalatinsk test site: A lot of investigations were devoted to chromosomal instability, which was established in persons living near the Semipalatinsk test site. The cytogenetic examination of 178 persons living in seven settlements of the Altai region exposed to ionizing radiation during the Semipalatinsk nuclear tests in 1949–1962, was carried out by V.A. Shevchenko et al. (1995). The frequency of chromosomal aberrations (dicentrics and rings) significantly exceeded the control level. The cells with more than one aberration were revealed in irradiated persons.

In 2003, Almaty scientists described chromosomal instabilities (Abil'dinova et al. 2003) in the persons who were born and have permanently lived in the contaminated zones of the Semipalatinsk region. A cytogenetic study has demonstrated that the frequency of aberrant cells is 1.7–3.0 times higher than control parameters. The total frequencies of chromosomal aberrations are 3.43 ± 0.48 and 1.15 ± 0.17 aberrations per 100 cells in the populations of the extreme radiation risk and control zones, respectively. The high chromosomal aberration rate in the radiation risk zones has been detected mainly due to radiation-induced chromosome markers, including dicentric and ring chromosomes and stable chromosomal aberrations. The data were published on the genetic polymorphisms and expression of minisatellite mutations in a 3-generation population around the Semipalatinsk test site (Bolegenova et al. 2009) as well as on the unstable-type chromosomal aberrations in lymphocytes from individuals living close to explosions (Tanaka et al. 2006).

Cited publications by V.Shevchenko and his colleagues, as well as Kazakh and Japanese scientists showed the total high chromosomal aberration rate, including the unstable and stable abnormalities in blood lymphocytes of persons living in the zone of the Semipalatinsk region.

6.1.2 Population of the Tundra Nenets Living in the Purov Region of the Yamalo-Nenets Nenets Autonomous Area as a Model to Investigate Consequences of Nuclear Test Fallout Impacts

For a long period of time, the Northern territory of West Siberia was being polluted by anthropogenic radionuclides from the fallout of the nuclear facilities at the polygons of the USSR, China, and the USA. The main contribution to the radiation-induced damage was caused by nuclear explosions in the Novaya Zemlya polygon: its energy-release quota consists of relatively 94 % of all nuclear tests in the USSR (Osipova et al. 2000a).

Long-term investigations of the Tundra Nenets population living in the basin of the Pour River (the Pour region of the Yamalo Nenets autonomous area) (Fig. 6.1) have revealed the increased frequency of gross CAs, derivative immunodeficient states and changes of the blood characteristics that are specific for inhabitants of the radiation-polluted regions. The long-lived radionuclides detected in the components of the biogeocenosis showed considerable modern radioactive pollution in the Northernmost inhabited territories of Russia. In this connection, the ecological food chain lichen-reindeer–man is very important. In 2000, the Cs^{137} contamination was averaged at 118.2 and 162.1 Bq/kg in lichen *Cladina stellaris* and venison,

6.1 General Genetic Consequences of Radiation Impacts

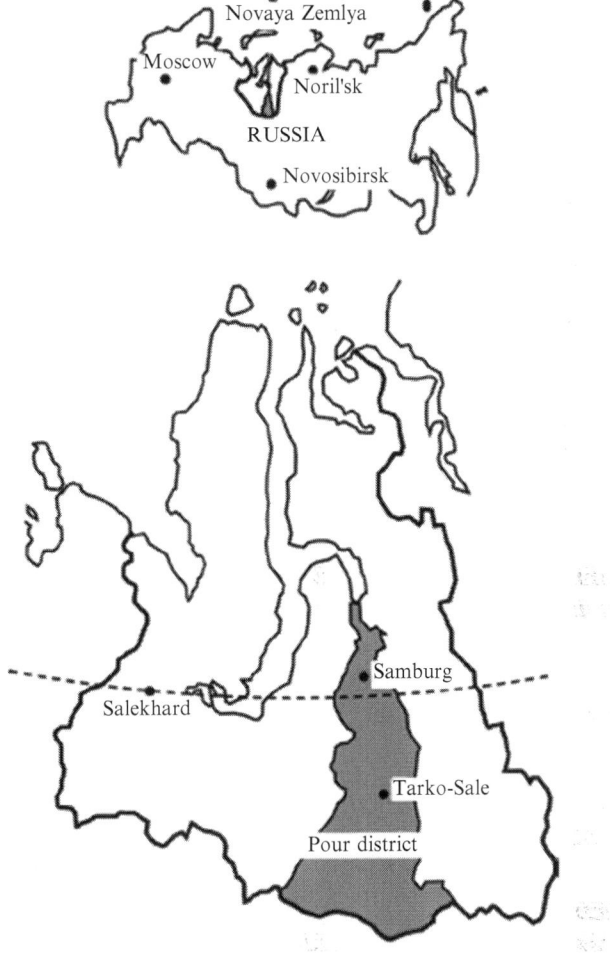

Fig. 6.1 Yamalo-Nenets Autonomous Area (Yamalia) (Republished from (Osipova et al. 2000a))

respectively.[1] Chronic consumption of this venison is a source of internal irradiation. Proximity of the tested territories to the Northern nuclear polygon allows one to consider it the main source of the radiation pollution after the nuclear tests in Novaya Zemlya (Osipova et al. 2000a).

Ninety individuals were randomly chosen, and the general analysis of their peripheral blood was carried out (Osipova et al. 2000b). The analysis showed that 17 % of the observed individuals had the maximal departure from the regional norm related to the decreased contents haemoglobin, leukocytes eosinophilia (it is not connected with helminthic invasion), and destruction of the erythrocyte form. Only 24 % of tested individuals had the normal blood characteristics.

---

[1] Studies were performed in the Novosibirsk Joint Institute for Geology, Geophysics and Mineralogy SD RAS.

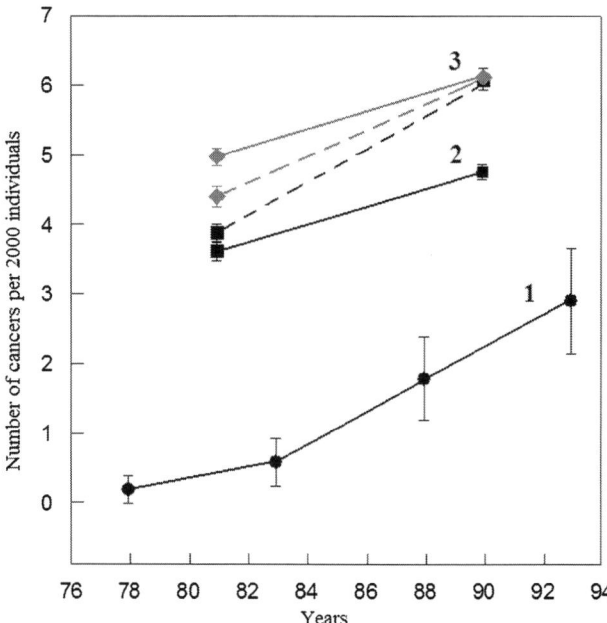

Fig. 6.2 Cancer mortality in the settlement of the Samburg population and the number of cancers in the Gomel and Mogilev (Belarus) inhabitants. The cancer mortality in the Samburg inhabitants (1976–1995) (*1, the black line*) (Osipova et al. 2000b). The std errors are shown (The high level of the std. errors is explained by the small numbers of the Tundra Nenets population (about 2,000 individuals)). The number of cancers in the Gomel (*2, the black lines*) and Mogilev (*3, the grey lines*) regions during 1977–1994 years (Yablokov et al. 2009); radionuclide pollutions: <5 Ci/km^2 (*the solid lines*) and >15 Ci/km^2 (*the dotted lines*)

Figure 6.2 shows the dramatically increased cancer mortality in this region, which had previously been unusual for the native population. It can be compared with the data on the number of cancers registered during the years 1977–1994 in the Gomel and Mogilev regions (Belarus) for low (<5 Ci/km^2) and high (>15 Ci/km^2) levels of radionuclide pollutions.[2]

To evaluate the introduction of radionuclides and their migration into the tissues of the native inhabitants (and the children especially), placentae samples of the Nenets (80) and Russian (10) women (1999–2000) were collected after they had given birth. The Russian women lived in the city of Novosibirsk. Radiocesium was not revealed in the placentae of the Novosibirsk Russian women. But ^{137}Cs was revealed in all placentae of the Nenets women, its content varied from 1 to 27 Bq/kg. So it has been established that the introduction of low irradiation doses into the tissues of the mother and child will negatively influence the health of both of them.

---

[2]The data on the number of cancers were given in (Yablokov et al. 2009).

The population of the Tundra Nenets living in the Pour region has been chosen as a model for complex genetic and ecological investigations, the reasons being as follows (Osipova et al. 2002):

1. This group lives relatively close to the Novaya Zemlya archipelago and the nuclear test polygon;
2. The small number of individuals (about 2,000 persons) and their ethnic homogeneity allow one to study little samples, limitation of their constant habitat, and migration beyond the tested region;
3. This group is well-studied from the point of view of its population and specific genetic features, providing regular genetic-ecological monitoring;
4. Preservation of its traditional seminomadic lifestyle and the simplification of study by the resultant basic food chain: lichen-reindeer-man, and water-fish-man.

6.1.3 Radiation Effects of the Semipalatinsk Nuclear Test Fallout in the South Baikal Zone (Pribaikal'e)

The soil contamination in the Altai and Baikal (Pribaikal'e) regions was studied only in the 1990s. At that time, ^{137}Cs-contamination was high in both regions and reached 200–400 Bq/kg.[3] Analyses showed that the Semipalatinsk nuclear tests (1949–1962) were the source of these pollutions (Boltneva et al. 1977; Scherbov et al. 2006) (Fig. 6.3).

The radiation fallout occurred in the 2–3 days after the nuclear tests, and their effects were serious: the accumulated doses of the external radiation exposure were on the average of 5–18 cSv (Scherbov et al. 2006; The Ministry of Nature of RF 1992). The summarized irradiation dose, which can be accounted for adult individuals, is 10–40 cSv; it is several times greater for the children (Nepomnyaschih et al. 1999). Both the total and infantile mortality and cancer levels in this region relate to the criteria of the ecological disaster area and the zone of the extreme ecological situation (The Ministry of Nature of RF 1992). Nowadays, the Pribaikal'e territory contamination is safe for human habitation and economic activity.

The settlement of Maloe Goloustnoe (Fig. 6.3) was chosen for the medicine and genetic investigations in the Irkutsk region. Maloe Goloustnoe is considered for the test site because:

1. It is located in the direction of the ground wind of the Semipalatinsk test site – Irkutsk in the 13.08.1953 and 25–26.08.1956, when the greatest radioactive fallout was observed in Irkutsk;

---

[3]The gamma-spectrometer analysis was performed in the laboratory of nuclear methods in the Central Analytic Laboratory "Sosnovgeologiya" (1992)

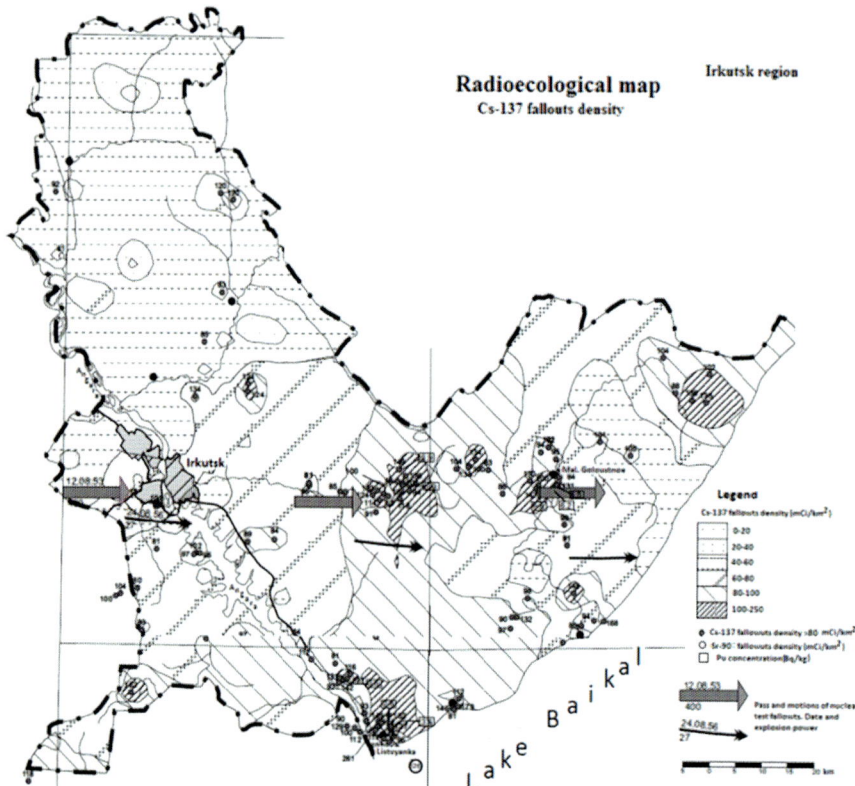

Fig. 6.3 The Irkusk region. Map of the ^{137}Cs fallout density (The map is made in "Sosnovgeos" on the data of "Sosnovgeosservis", the Institute of Geochemistry SD RAS, and "Irkutskgeologiya". The authors are L.G. Korshunov, V.I. Medvedev (Medvedev et al. 2005).)

2. It is characterized by a high level of anthropogenic radioactive pollution: ^{137}Cs settling density on soil 200 mCi/km^2 and Pu contamination 4.5 ÷ 6.8 Bq/kg;
3. It is a safe settlement in regard to the chemical pollution that makes it possible to evaluate the influence of the radiation factor "per se." The medical-genetic and cytogenetic investigations showed the steady health impairment across generations that verifies late consequences of nuclear tests and genome instability (Medvedev et al. 2005, 2009).

A group of individuals living in the settlement of Listvyanka was analyzed in addition to the Maloe Goloustnoe sample. Listvyanka experienced radiation impacts due to the Semipalatinsk explosions and its location in the polluted area (Medvedev et al. 2009).

6.2 Objects and Methods of Radiation Epidemiology Investigations

6.2.1 Description of the Tested Sites and Populations in the Yamalo Nenets Autonomous Area

Characteristics of the polar Tundra Nenets population living in the settlement of Samburg and neighboring tundra: The population of Tundra Nenets living in the settlement of Samburg and neighboring tundra (the Pour region of the Yamalo-Nenets autonomous area) is about 2,000 individuals who are settling in an area of approximately 100,000 km² and have a mainly nomadic life (Fig. 6.4). This population is ethnically homogeneous, and the migration quota from without and into this area is small. The economy is based on reindeer breeding, fishing, and hunting. The population is genetically well-studied. The samples of tested individuals consisted of those who had immediately experienced the radiation impacts and their progeny.

Radiation situation in the Pour region: At the end of the previous century, the components of biogeocenosis were analyzed: forest floors, mosses, lichens, peats, and sea-floor sediments, and also some samples of vegetation that are food for the reindeer in summer. Samples of venison, which is the main food for Tundra Nenets,

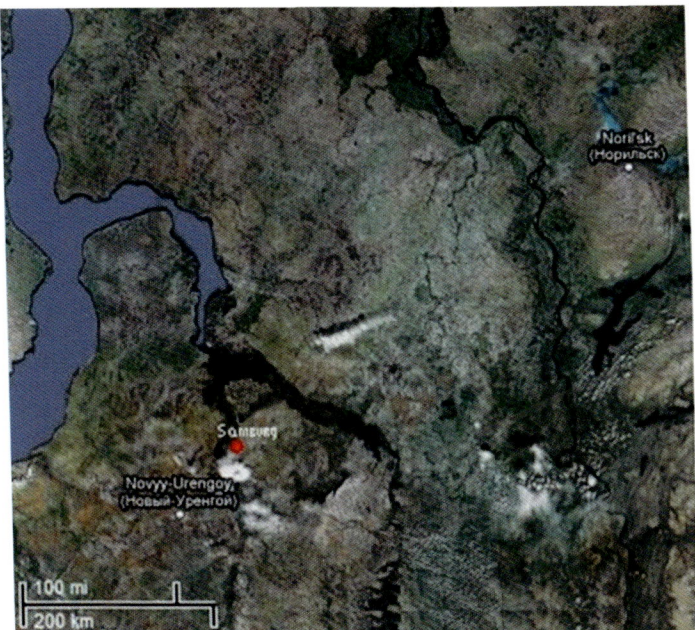

Fig. 6.4 Map of the Pour region of the Yamalo-Nenets Autonomous area

Table 6.2 ^{137}Cs – activity in the components of the biogeocenosis of the Pour region (1998–2000)[a]

Components	^{137}Cs, Bq/kg	Components	^{137}Cs, Bq/kg
Forestfloors	320 (19–610)	Liver	53 (17–131)
Mosses	89 (2–738)	Kidneys	140 (46–317)
Lichen	118 (9–372)	Lungs	54 (30–82)
Peats	130 (60–200)	Heart	65 (47–99)
Sea-floor sediments	59 (33–102)	Bone marrow	15 (2–38)
Fungi white dry	194 (170–219)	Bone tissue	18 (3–57)
Leaf of alder and birch	64	Fish	6 (0–10)
Horsetails	42 (8–76)	Cowberry	7 (2–12)
Venison	213 (48–315)	Cranberry	10
Dried venison	1,200		

The data are given in terms of air-dry mass excluding the samples of different reindeer organs, which were analyzed in the native state. The interval of the ^{137}Cs contamination is given in the brackets

[a]These data were presented by the Institute of Cytology and Genetics SD RAS (Novosibirsk), the Joint Institute for Geology, Geophysics and Mineralogy SD RAS (Novosibirsk), and the Institute of Chemical Kinetics and Burning SD RAS (Novosibirsk)

were collected. ^{137}Cs was determined by the gamma-spectrometric, and ^{90}Sr – beta-radiometric methods at the Joint Institute for Geology, Geophysics and Mineralogy of the Siberian Department of the Russian Academy of Sciences (Novosibirsk). The results of the ^{137}Cs – activity are shown in Table 6.2.

Many publications are devoted to the problem of "cup moss-reindeer-man", but these investigations date back to the 1960s–1970s (Liden 1961; Nizhnikov et al. 1969). Nowadays, ^{137}Cs is detected in all tested sites; its activity for the biogeocenosis components varies widely (Table 6.2) (Osipova et al. 2000a, b).

Undoubtedly, these variations are, first of all, caused by irregular radioactive fallout that was mentioned by all radioecologists, although the species of plants is of great importance. So, the spread of the ^{137}Cs activity is 58–144 Bq/kg for the different species of lichen collected in the area of 10 m × 10 m in the Pour region.

The ^{137}Cs activity in venison is high enough (Table 6.2) to be a constant source of chronic internal radionuclide irradiation for the Tundra Nenets. The dried venison is a significant part of the summer ration of the native inhabitants and the value of 1,200 Bq/kg per one sample of the dried venison is very dangerous. For comparison: nowadays, ^{137}Cs is not detected in mutton or beef from the Altai territory which is located near the Semipalatinsk test site.

As for ^{90}Sr, its contamination in the reindeer bone tissue is high enough and varies within the 108–555 Bq/kg contamination that exceeds the overload capacity (200 Bq/kg) in some samples.

The long-term complex investigations showed that the territory of the Pour region of the Yamalo-Nenets autonomous area experiences the anthropogenic factor, and the residual radioactive pollution has a dominant role in comparison with pollutions by heavy metal and fallout of the oil and gas complex (Osipova et al. 2000a, b).

Table 6.3 Cells with chromosomal abnormalities (CCAs) in the population of the Tundra Nenets (Osipova et al. 2000a)

Group	Number of individuals	Number of metaphases	Total number, occurrence frequency and confidence interval	
			Aberrant cells	Cells with rings and dicentrics
Total	170	18,406	588	83
			3.19	0.45
			(2.9–3.4)	(0.36–0.55)
Adult individuals	114	12,310	444	70
			3.61	0.57
			(3.3–3.9)	(0.44–0.71)
Children	56	6,096	144	13
			2.36	0.21
			(2.0–2.8)	(0.11–0.34)

The individuals living in the tested site who experienced the radiation impacts or their offspring were investigated. The table was taken from the paper (Osipova et al. 2000a) The worldwide data can be used as the control: total number of aberrant cells is 0–1.5 %, number of cells with rings and dicentrics is 0.05–0.21 % (Bochkov 1993)

Genetic and medical investigations of inhabitants of the Siberian North tundra:
The genetic investigations performed in 2000 showed the total high frequency of chromosomal aberrations and significant occurrence of the radiation markers (Table 6.3). It was revealed that the occurrence frequencies of the total CAs and both rings and dicentrics are 3.19 and 0.45 % that exceeds significantly the upper limits of the accepted norms of these values equal to 1.5 and 0.1 %, respectively (Osipova et al. 2000a). For adult individuals, these values are higher: 3.61 and 0.57 %, respectively.

In the children's group (N = 56), the general frequency of cells with abnormalities is equal to 2.36 %, which exceeds significantly ($p < 0.05$) the control level. The rings and dicentrics are revealed in the blood of 13 children from 56 (this is 23 % from the total childrens' sample). The average frequency of the radiation markers observed in the blood of the Samburg children is 0.21 %, which exceeds the upper limit of the datum-control-level by two times. The same cytogenetic investigations were carried out in the children with multiple defects of development and atypical jaundice of the newborn living together with their parents near the Semipalatinsk test site (Matveeva et al. 1993). It was shown that the general frequency of the cells with abnormalities in these children was 2.8 and 2.7 %, respectively, and 40 % of the total aberrations are of the chromosomal type (breaks, changes, rings). So the level of the CAs in the Samburg children is comparable to the same in the children with multiple defects of development and atypical jaundice of the newborn living in the Altai radiation-polluted zone. Although the sample size is not large, the obtained data can indicate both transgeneration of the genetic abnormalities and existence of the current strong ecological factors which influence the stability of the chromosomal apparatus in cells.

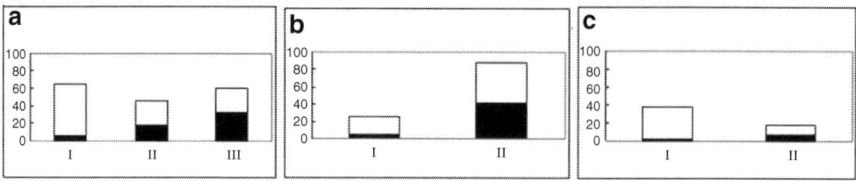

Fig. 6.5 Distribution of the tested samples on groups with the lower (*I*), moderate increase (*II*) and high general (*III*) frequency of cells with abnormalities. The total sample of Tundra Nenets (170) (**a**), adult part of the sample (114) (**b**), children (56) (**c**), x-axis: the groups of tested individuals; y-axis: number of the individuals. Quota of the individuals with radiation markers (rings and dicentrics) is *black colored*

The total sample (N = 170) of individuals was divided into three groups with different levels of the general frequency of CCAs (Osipova et al. 2000a): I – the frequency of CCAs is lower than the control one (the frequency of CCAs $\leq$ 1.5 %), II – moderate increase of the frequency of CCAs (1.5 % < the frequency of CCAs $\leq$ 3 %), and III – high frequency of CCAs (the frequency of CCAs > 3 %) (Fig. 6.5). It is shown that the frequency of CCAs does not exceed the control in 38% of individuals; it moderately increases and exceeds the control by more than two times in 27 and 35 % of the individuals, respectively. The radiation markers – the rings and dicentrics – were observed in 9, 39, and 53 % of individuals in the first, second and third groups. The frequency of the radiation markers tends to increase with the frequency of cells with general CAs.

The authors (Osipova et al. 2000a) assumed a positive dynamics for the children's group (C): the group for which the frequency of CCAs approximates the control level (I) is two times bigger than the group with the increased frequency of CCAs (II).

The data on the number of cells with abnormalities is presented for samples of the individuals living in the settlement of Samburg and Polar Tundra in Sect. 8.6.1.

In 2002, studies were performed to compare the aborigines' and Caucasoid migrants' data (Osipova et al. 2002). The results are shown in Table 6.4, and generally they are in agreement with the data presented in Table 6.3: the elevated frequencies of CCAs are registered in blood lymphocytes of the Samburg Tundra Nenets, and only in 25–30 % of the examined grown-ups do the CCA levels not exceed the control group. The situation is somewhat better with the subgroup of Nenets children, where about 2/3 have normal levels. If Nenets or migrants have elevated CA levels in their blood, then there is a tendency to accumulate radiation markers. The cytogenetic data on the migrants differed surprisingly strongly from the norm. This fact may be explained by either – their having lived in the ecologically hostile environment of the North for a long time or by the year or place of their birth (Chelyabinsk, Perm, etc.).

6.2 Objects and Methods of Radiation Epidemiology Investigations

Table 6.4 Cells with abnormalities in the Samburg Tundra Nenets vs migrants

Sample size	N	Metaphases, number	Aberrant cells, total, %; confidence interval	Cells with rings and dicentrics, %; confidence interval
Total	246	26,808	821	113
			3.06 (2.9–3.3)	0.42 (0.35–0.5)
I. Aborigine	171	18,436	587	77
			3.18 (2.9–3.4)	0.42 (0.35–0.52)
II. Adult Caucasoid migrants	75	8,372	234	36
			2.8 (2.4–3.1)	0.43 (0.3–0.57)
Adult aborigine	114	12,240	442	65
			3.61 (3.3–3.9)	0.53 (0.41–0.67)
Aboriginal children born in 1980 or later	57	6,196	145	12
			2.34 (1.9–2.7)	*0.19 (0.1–0.31)

The data statistically ($p < 0.001$) exceed the control levels in all groups except where asterisked ($p > 0.05$). Data was published in (Osipova et al. 2002)

The worldwide data were used as the control: the total number of aberrant cells is 0–1.5 %, number of cells with rings and dicentrics is 0.05–0.21 % (Bochkov 1993)

Note that 11 % of the blood samples failed to grow on the culture medium. It is possible that this is due to the so-called "immune depression," since secondary immunodeficiencies were not rare in those people.

The immunologic investigation was carried out in the population of Tundra Nenets showing the specificity of their immune status (Osipova et al. 1999): the secondary immunodeficiency states were observed in 27 % of the Tundra Nenets whereas the percentage of the healthy ones was 23 (in the town of Tyumen, 14 and 3 %, respectively). The number who suffered from chronic inflammatory diseases increases among the aged individuals.

The comparison of these investigations with the results of the cytogenetic studies of the blood lymphocytes has shown that the quota of individuals who suffered from secondary immunodeficiency is high in the group of individuals whose blood lymphocytes are not stimulated, giving evidence of the so-called "immune depression" of cells, and the group of individuals with secondary immunodeficiency is characterized by a higher (1.7 times) frequency of cells with radiation markers (rings and dicentrics).

Apparently, radiation is a crucial factor for chromosomal instability. These data are of a special prognostic importance because they indicate higher risks of developing pathologies, including cancer diseases. Thus, the parents who were born in 1965 and 1969, and had rings and dicentrics in their blood, gave birth to a blind daughter in 1997. Cancer accounts for several deaths in the Nenets between 1992 and 2001, and all cases had cytogenetic figures considerably beyond the norm.

Fig. 6.6 Lake Baikal, Irkutsk region. Settlements Maloe Goloustnoe and Listvyanka (signed by *red color*)

6.2.2 Description of the Tested Sites and Populations in the Pribaikal'e Region

The human population living in the settlements of Maloe Goloustnoe and Listvyanka: The settlements of Maloe Goloustnoe and Listvyanka are located in the Irkutsk region (Fig. 6.6), and are mainly populated by Russians. The soils are used for agriculture. Listvyanka is the most popular place at the Baikal Lake, and a major tourist center of the Baikal today.

Radiation situation in the Pribaikal'e tested sites: During the nuclear test operation in the Semipalatinsk polygon (1949–1962), the radiation effect on the persons living in South Pribaikal'e was considerable (Medvedev et al. 2005, 2009). Figure 6.3 shows the trajectory of the air streams spreading the nuclear test products from the Semipalatinsk test site resulting in strong fallout in the South Pribaikal'e region (1953, 1956). The trajectory (12.08. 1953) passed through Maloe Goloustnoe. High density of the ^{137}Cs is observed in Listvyanka, where the ^{137}Cs fallout density exceeded 80 mCi/km^2. For the Maloe Goloustnoe and Listvyanka sites, the accumulated doses were of 5–18 cSv due to external irradiation (The Ministry of Nature of RF 1992). These doses could increase up to 10–40 cSv for adults, and they were sometimes higher for children (Nepomnyaschih et al. 1999) if we take into account the internal J^{131} irradiation with home-produced milk.

Investigations of inhabitants of Maloe Goloustnoe: The investigations were performed with the individuals living permanently in Maloe Goloustnoe (2003–2006) (Osipova et al. 2008). The tested sample consisted of individuals born in 1950–1989, and they or their parents join the risk group which had experienced the fallout of the Semipalatinsk nuclear test (12.08.1953). The cytogenetic sample consisted of 64 individuals from the 243 people tested. The total frequency of cells with abnormalities was observed, and it was shown that for 10 individuals (that is 16 % of the tested sample) the frequency of CCAs was approximately equal to the control level (1.5 %). The cytostatic and cytotoxic effects were detected in seven individuals (it is 11 %). Specific radiation markers were revealed in 24 individuals (43 %) who often had other types of CAs, such as inversions and translocations. The multi-aberrant cells (rogue cells) were revealed in ten individuals, and were always accompanied by other types of abnormalities: rings, dicentrics, translocations, cytostatic and cytotoxic effects. The frequency of cells with all types of abnormalities was 4.3 %. Besides them, the aneuploid cells were observed (47, XXX; 48,XXXX; 47,XX + 21), and the frequency of the aneuploidy was 0.2 %. These results have demonstrated considerable cytogenetic effects in the risk group of individuals living in Maloe Goloustnoe. The observed high frequency of the multi-aberrant cells and unstable chromosomal aberrations (rings and dicentrics) suggest that the leading part of these effects belongs to the ionizing irradiation caused by the Semipalatinsk nuclear test.

6.2.3 Cytogenetic Analysis

The Novosibirsk Institute of Cytology and Genetics of SD RAS performs regular investigations of blood lymphocytes in the Tundra Nenets population and longtime inhabitants of the Irkutsk region. The data used for statistical modelling were collected during many years of studying.

The material for the cytogenetic analysis was peripheral blood lymphocytes of the individuals. Lymphocyte cultivation was carried out by using the standard macromethod (Moorhead et al. 1960). Fixation of the cultures was provided for 48 h of cultivation (the first cell division). The total spectrum of the structural aberrations of the chromatid as well as of chromosome types was considered to be a criterion for assessment of the mutagenic effect (Cao et al. 1981). The so-called "gaps" and aneuploid cells were not taken into consideration. Identification of the nature of chromosome damage and detection of the mutations of some karyotypes was made on differential colored preparations (Seabright 1971).

Frequency of lymphocyte cells with abnormalities was determined as the ratio of the abnormal lymphocytes number to the proliferated ones. The results of the analyses of chromosomal breakages in blood lymphocytes of the individuals living in these settlements are published in (Osipova et al. 1999, 2002).

6.3 Statistical Modelling for Persons Living in the Radiation – Polluted Areas

6.3.1 Methods of the Statistical Modelling

Modelling: In the case of blood lymphocytes, we need to address the frequency of cells with abnormalities. One should analyze the distributions of individuals on the occurrence frequency of abnormal cells among proliferated cells (PCs). This issue was specially investigated because the numbers of both abnormal cells and PCs are independent values (see Sect. 3.4.4). It has been shown that the distributions on the frequency of cells with abnormalities can be used to analyze instability processes (Florko et al. 2009), but individual distribution on the number of PCs should be accounted. The individual numbers with a small number of PCs influence the distribution tail.

Samples of individuals: Random samples of the persons living in the settlements of Samburg and Maloe Goloustnoe were studied. The samples of persons were divided into four groups corresponding to the individuals who had been immediately irradiated by the fallout, and their children, grandchildren, and great-grandchildren. The aged individuals, who had experienced the radiation fallout in the case of the Listvyanka site, were investigated.

The sample of known healthy individuals was used as the control group. They are the individuals living in the city of Novosibirsk because it is difficult to find an ecologically pure region in Siberia. The blood sample (1 ml) of the individuals from the control group should contain not less than 100 activated lymphocytes.

Approximations: The methods of approximation, assessment of its efficiency, and statistical treatment were the same as for investigations of the appearance of cells with abnormalities in plant seedling meristem (see Sect. 8.1). The best hypothesis with respect to the majority of criteria was preferred. For close values of criteria, the "simpler" hypothesis was chosen.

As a rule, the statistic samples are small in the case of modelling for persons. At present, the modern special mathematical methods have been worked out to study the small samples. Some of these are presented in Sect. 8.1. The small-sample statistics were used; therefore, the stability of distributions was always verified. The verification consisted in variation of the length of the partitioning interval for histogramming.

6.3.2 Examination of the Distributions of the Control Group of Individuals

The control group consists of the cytogenetic data provided by the Novosibirsk Institute of Cytology and Genetics of SD RAS and the town medical laboratories.

6.3 Statistical Modelling for Persons Living in the Radiation – Polluted Areas   123

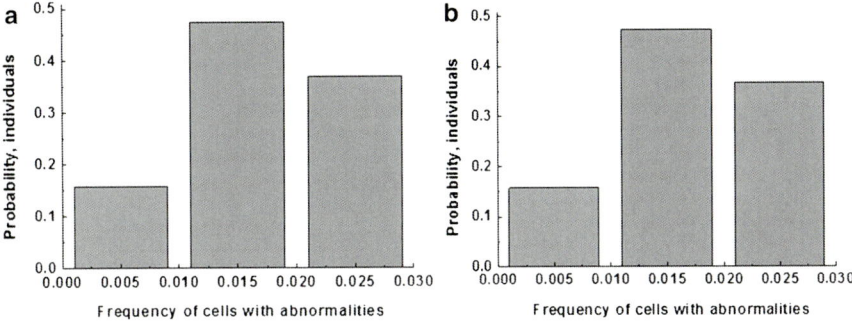

Fig. 6.7 Comparison of the distributions of individuals on the frequency of cells with abnormalities. Age of the individuals is <25 (**a**) and 25–50 years (**b**)

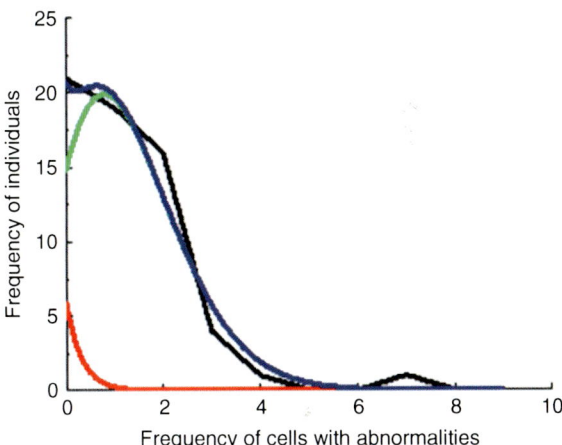

Fig. 6.8 Distribution of the frequency of individuals on the frequency of cells with abnormalities. The modelling was performed on the scientific laboratory data (accounted 100 metaphases) jointly with "medical" data (selected N metaphases ≥ 30). Experimental data – *black*, Poisson – *green*, geometric – *red*, and sum G+P is a *blue color*. Poisson distribution: 90 %, sample mean 1.3; geometric – 10 %, sample mean 1.0

To be sure in the control, both groups of data were examined for their homogeneity. In the first case, the identity of the distributions for different ages (<25 and 25–50 years) was checked (Fig. 6.7). The second sample was divided into sub-samples corresponding to different numbers of analyzed metaphases. The first data showed full identity of the distributions for different ages.

Examination of the medical data showed their unsuitability for the investigations (see Sect. 8.6.2), mainly because the first 12–13 metaphases had been analyzed. A small part of these medical data was used to estimate the "control" distribution of the individuals on the frequency of cells with abnormalities (Fig. 6.8). The control group was completed by the cytogenetic analyses where the number of the checked metaphases exceeded 30. The data performed in the scientific laboratory were analyzed for the final conclusions.

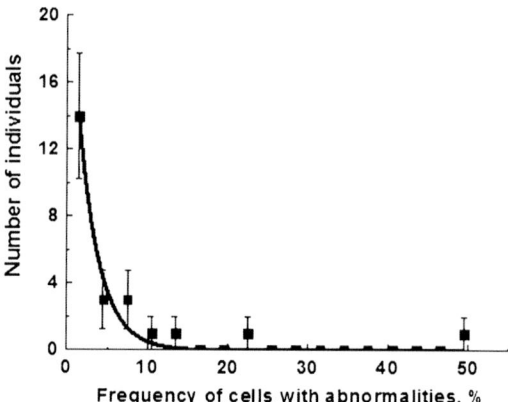

Fig. 6.9 Experimental distribution of individuals on the frequency of cells with abnormalities and its approximation by the geometric law. The sample of individuals living in the settlements of Listvyanka, Maloe Goloustnoe, Samburg, and South of the Pour region. Number of accounted lymphocytes was $N_{mph} < 30$. Experimental distribution – the points with the std. errors. Modelling by the geometric law – *the line*, the sample mean $mG = 2.87$. Model efficiency: $\chi^2/df = 0.28$, $df = 16$, $p > 0.01$

6.3.3 Comparison of the Distributions of Individuals with Normal and Low Activated Cells on the Frequency of Cells with Abnormalities in Blood Lymphocytes

Medical-genetic investigations of Tundra Nenets population revealed connections of high secondary immunodeficiency of individuals with low activation of their blood lymphocytes ("immune depression" of cells) (Osipova et al. 1999). Moreover, the cytogenetic analysis has shown high frequency of cells with abnormalities, multiple aberrations (rock cells) and high frequency of cells with radiation markers (rings and dicentric) in these cases.

It is of interest what type of distribution corresponds to the case of bad activated lymphocyte cells. The samples of individuals were chosen to compare the distributions for individuals with bad and normal activated blood lymphocytes. These samples consisted of individuals whose number of accounted lymphocytes was less than 30 $N_{mph} < 30$ and equal to 100 $N_{mph} = 100$ (city of Novosibirsk, laboratory data). The first sample was chosen from the individuals living in the Irkutsk region (the settlements of Listvyanka and Maloe Goloustnoe), and Yamalo-Nenets autonomous area (the settlement of Samburg, South of the Pour region).

Experimental distributions for the individuals whose number of accounted lymphocytes was less than 30 $N_{mph} < 30$ and equal to 100 $N_{mph} = 100$ on the frequency of CAs, are shown in Figs. 6.9 and 6.10. The statistical modelling shows that the more effective models are the geometric for individuals whose number of accounted lymphocytes was less than 30 $N_{mph} < 30$ (Fig. 6.9) and the Poisson in the case of accounted lymphocytes equal $N_{mph} = 100$ (Fig. 6.10). It assumes that the types of the distributions of the individuals with normal and bad activated blood lymphocytes are different.

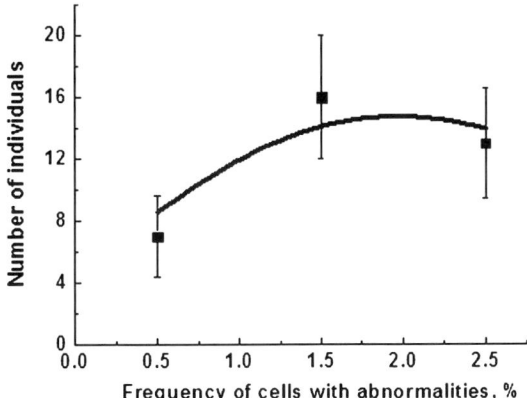

Fig. 6.10 Experimental distribution of individuals on the frequency of cells with abnormalities and its approximation by the Poisson law. Control sample of the individuals living in the city of Novosibirsk. Number of accounted lymphocytes was $N_{mph} = 100$. Experimental distribution – the points with the std. errors. Modelling by the Poisson law – *the line*, the sample mean $mP = 1.65$. Model efficiency: $\chi^2/df = 0.32$, $df = 2$, $p > 0.01$

Table 6.5 Number of individuals whose blood cells were bad or not activated (the Samburg sample, 1997)

	n_{total}	n_1 ($N_{mph} < 30$)	n_2 ($N_{mph} = 0$)
Number of individuals	163	13	23

The sample means of the P- and G-laws are $mP = 1.65$ and $mG = 2.87$, respectively. The P- sample mean is low, and we can suggest the spontaneous nature of abnormalities (this parameter should be increased when cells with abnormalities have accumulated, see Sect. 3.4.1). It is expected for the normal activated blood lymphocytes and healthy individuals.

The geometric law is revealed for individuals with poorly activated blood lymphocytes. We can suggest the domination of the instability together with the selection processes in the blood of this group of individuals. It is understandable because the instability and selection processes mean a growing number of dead sensitive blood cells that leads to depletion of the lymphocyte pool. On the other hand, the cells' instability can be closely related with the bad activation of lymphocytes. All these effects result in a high percentage of fatal cases of sensitive individuals. The medical examination showed that individuals with bad activated blood cells had suffered different diseases including the blood illness (see Sect. 6.2.1).

We can alternatively check the assumption that the data with the decreased number of activated lymphocytes are related to the cases of poorly stimulated blood cells. This situation is observed in the Samburg sample (Antonova et al. 2008a): the maximal number of individuals with a minor number of active accounted blood cells (n_1) corresponds to the maximal number of individuals without the growth of stimulated cells (n_2) (Table 6.5). In Sect. 6.3.5, we show that the Samburg sample of individuals is geometrically-distributed.

6.3.4 Comparison of the Distribution Structures of Individuals Under 18 and Those Older Living in the Pribaikal'e Region on the Frequency of Cells with Abnormalities

The samples of individuals living in the settlements of Maloe Goloustnoe and Listvyanka are characterized by high frequency of cells with abnormalities in blood lymphocytes (Osipova et al. 2008; Medvedev et al. 2009). The examination was performed with these data on the homogeneity by comparison of the distributions for men and women (see Sect. 8.6.3) to be sure of their high quality.

The structure of distributions of these individuals was studied to investigate processes across generations related to these high frequencies (Antonova et al. 2008b).

The sample of individuals living in Maloe Goloustnoe was divided into four age groups: parents (60–80), children (40–59), grandchildren (18–39), and great-grandchildren (<18). The sample of individuals living in Listvyanka (23 individuals) consists of the "parents".

The modelling was performed (see Sect. 8.6.4), and the best distributions are shown in Fig. 6.11. The type of distribution for samples of individuals living in Maloe Goloustnoe changes between generations:

$$G \to G + P \to P \to P \qquad (6.1)$$

We see a tendency to decrease the relative value of the geometric component and the sample mean of the Poisson distribution, although the Poisson sample mean for the great-grandchildren's group preserves its higher value (3.1) in comparison with the control group (the city of Novosibirsk) (1.4).

We can assume that the geometric distribution means instability and selection processes which can lead to the death of individuals with multiple numbers of cells with abnormalities (Fig. 6.11a, b). The youngest generation groups are the Poisson-distributed with increased sample means (Fig. 6.11c, d) that assumes accumulation of abnormalities (see Sect. 3.4.1).

So, we can suggest the transgeneration of abnormalities. The authors (Osipova et al. 2008) pointed out the leading part of unstable aberrations – radiation markers rings and dicentrics – in the increased total frequency of abnormalities in the group of individuals 19–50 years old (they are "children" and "grandchildren"). There are a number of individuals with a small number of activated PCs in the samples of parents (a), children (b) and grandchildren (c). It was shown that distribution tails become longer in this case (see Sect. 3.4.4, Fig. 3.16).

What is the reduction of the geometric component between generations? It is known that aging induces free radicals in cells (Pinzino et al. 1999) which can activate cell elimination mechanisms (Davis et al. 2001), and the reactive oxidative species are involved in the stress-induced bystander mechanism (Averbeck 2010). So, the age influences the processes, which are described by the geometric law. This regularity was observed in the lab experiments and modelling for old pea seeds

6.3 Statistical Modelling for Persons Living in the Radiation – Polluted Areas 127

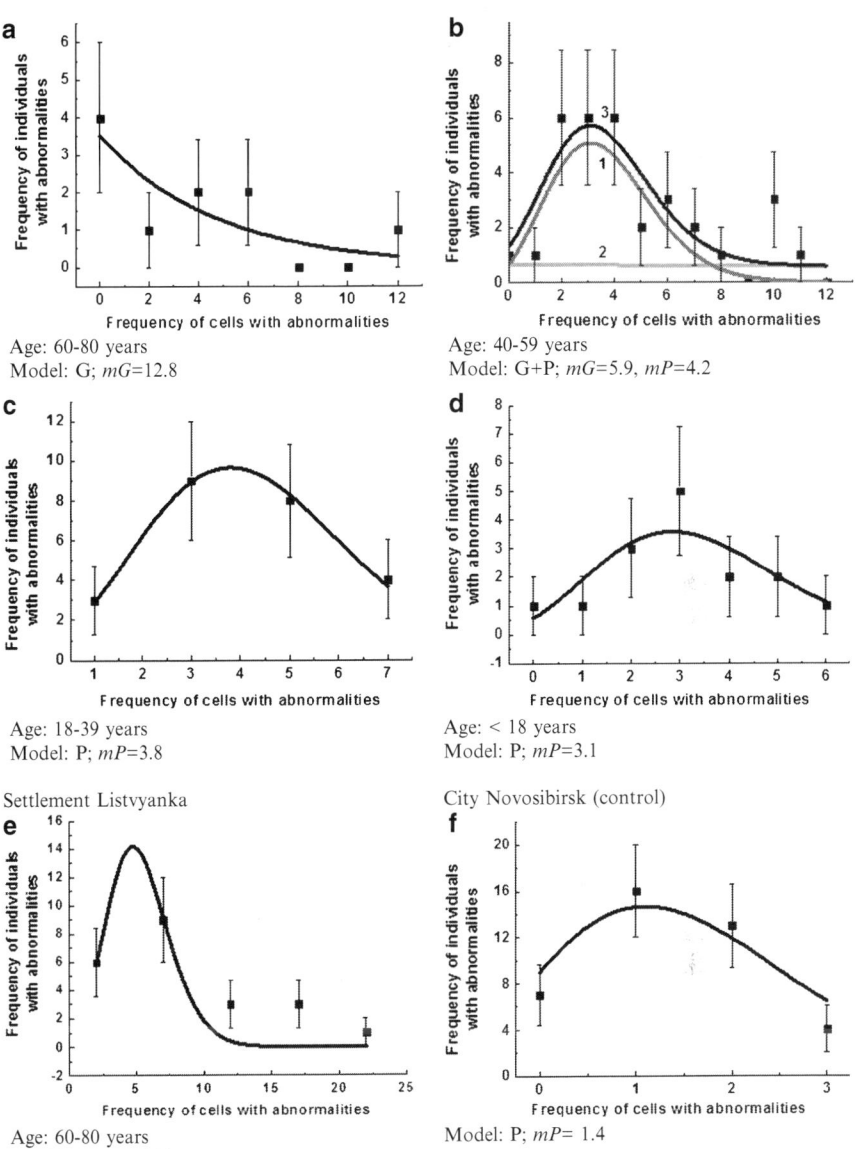

Fig. 6.11 Approximations of the experimental distributions of individuals' samples on the frequency of abnormalities (that is the cells with abnormalities). Standard errors are shown for experimental data. The model plots: (**a**) G-line; (**b**) P (1), G (2), sum G + P (3) lines; (**c–f**): P-line. Standard errors of the sample means: $d(mP) = 10\text{–}15\,\%$; $d(mG) = 30\text{–}40\,\%$; $d(mP) = 10\text{–}15\,\%$

Table 6.6 Types of distributions and their parameters for samples of the individuals of different ages living in the settlements of Maloe Goloustnoe, Listvyanka and the city of Novosibirsk

Settlement	Age	Number of individuals	Sample means of the distributions	
			G (mG)	P (mP)
Maloe Goloustnoe	60–80	11	3.0	
	40–59	30	5.8	4.2
	18–39	23		4.0
	<18	15		3.1
Listvyanka		23		4.8
Novosibirsk		40		1.4

Standard errors of the sample means: $d(mP) = 10\text{–}15\ \%$, $d(mG) = 30\text{–}40\ \%$

(see Table 4.4): this group of seeds is related to the increased P- and G-sample means and significant decreasing of seed survival. The tendency to decrease the P-sample mean with the generation (Table 6.6) could be clarified in terms of aging. But it does not exclude a relaxation of the transgenerational process.

The Listvyanka sample ("parents") is described by the Poisson law (see Sect. 8.6.4):

$$P \qquad (6.2)$$

There are individuals with a small number of PCs in the Listvyanka sample that provide a long tail of the Listvyanka P-distribution (see Sect. 8.6.4).

In these studies, we see different consequences of the Semipalatinsk nuclear tests for inhabitants of Maloe Goloustnoe and Listvyanka. The samples of persons from these two settlements are described by means of the G, G + P and P models. These models are related to different genetic processes in the blood of individuals.

The model $G \to G + P \to P \to P$ describes the processes of instability and selection, and is related to the Maloe Goloustnoe site. These radiation-induced processes continue in the next generation. We can assume that they relax because the geometric component value and the sample mean of the Poisson decrease with generations. The other explanation is decreasing of the bystander processes in young groups. In any case, the P-sample mean is elevated for the great-grandchildren of the persons who experienced directly the fallout impact that can reflect the late processes.

We can also think that the radiation effects for the inhabitants of Maloe Goloustnoe were stronger than for the individuals living in Listvyanka. It results from the P- distribution of the Listvyanka "parents," although its sample mean is greater than in the Maloe Goloustnoe P-sample means.

6.3.5 Analysis of Distributions of the Individuals Living in the Siberian Utmost Northern and Pribaikal'e Regions

The setup of the Tundra Nenets is very specific due to the frozen soil. The Northern food chain is lichen-reindeer-man. Lichens have been widely investigated

6.3 Statistical Modelling for Persons Living in the Radiation – Polluted Areas 129

for their potential accumulation of numerous stable and radioactive elements; they are suitable as quantitative biomonitors for atmospheric emissions of tritium and radiocarbon from nuclear facilities (Daillant et al. 2004). We can compare radiation effects of nuclear tests in Novaya Zemlya on native inhabitants of the Siberian North with the consequences of the Semipalatinsk nuclear tests on the Pribaikal'e inhabitants.

Let us analyze the distributions of individuals on the occurrence frequency of lymphocytes cells with abnormalities in blood samples of individuals living in the settlements of Samburg (North Siberia) and Maloe Goloustnoe (Pribaikal'e). The control group consists of healthy persons living in the city of Novosibirsk. The samples were divided into groups of individuals who were immediately irradiated by the nuclear test fallout and their children, grandchildren, and great-grandchildren. The experimental distributions and their approximations are demonstrated in Fig. 6.12. Table 6.7 presents the frequency of the abnormal cells occurrence in blood lymphocytes of individuals living in the tested sites and the city of Novosibirsk, and parameters of the most effective model of approximation.

The best "North" approximations can be presented by the series:

$$G + P \to G + P \to G + P \to P \qquad (6.3)$$

Decreasing of the instability process in the next generation: The "G + P" model relates to the processes of instability and selection, and can be used to describe the radiation effects for both the Tyumen and Irkutsk populations. In both cases, the geometric component is replaced gradually by Poisson, and the total samples of grandchildren (Irkusk region) and great-grandchildren (the Irkutsk and Tyumen regions) are the Poisson-distributed (1, 3). A tendency has been observed towards the decreased sample means of the P distribution across generations ($4.2 \to 3.1$), but it preserves a higher value ($p < 0.05$) than the control group (1.4) (Table 6.7). This tendency to relax may reflect either a decrease in the instability and selection process across generations, or a reduction of the reactive oxidative species within a generation which influences the bystander processes. Cytogenetic studies also showed a tendency towards the decrease of the averaged frequencies of lymphocyte cells with abnormalities within a generation for the Tyumen region (Fig. 6.5).

Difference between the consequences of the nuclear tests for the North and Pribaikal'e populations: There is a difference between the distribution types, which describe the genetic processes in the settlements of Samburg and Maloe Goloustnoe (series 6.3, 6.1) (Table 6.7). It means different radiation impacts (see Sects. 3.4.1 and 3.4.2). We can assume that the Novaya Zemlya fallout were stronger, and the situation can be worsening due to the long storage of radioactivity in lichens, mosses and forest floors. The prolonged radiation influence is verified by the absence of lymphocyte activation in the third elder Samburg groups (Table 6.7).

The sample mean of the P-distribution for both the Tyumen and Irkutsk regions varies a little, but significantly exceeds the control value. It means that the instability process continues across three generations of irradiated persons (Fig. 6.12). The authors (Osipova et al. 2000a) have concluded that the frequency of cells with

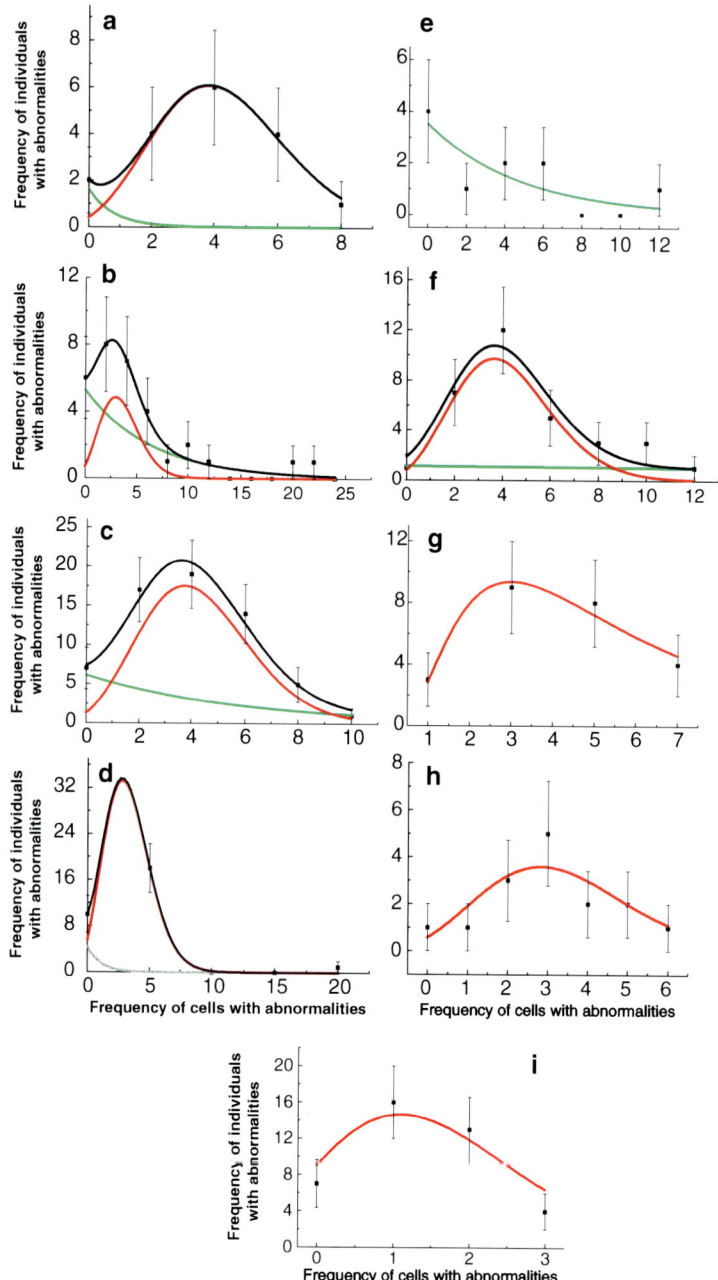

Fig. 6.12 Statistical modelling of distributions of persons of different generations on the frequency of the occurrence of CCAs (Korogodina et al. 2010b). Sample of the persons living in settl.Samburg (**a–d**); settl. M. Goloustnoe (**e–h**); (**c**) Novosibirsk (I). Parents (**a, e**); children (**b, f**); grandchildren (**c, g**); great-grandchildren (**d, h**).Total curve of approximation ___; Poisson component ___; geometric one ___. Standard errors of the total approximation curve are shown

6.3 Statistical Modelling for Persons Living in the Radiation – Polluted Areas

Table 6.7 Frequency of cells with abnormalities in blood lymphocytes of persons living in the Tyumen and Irkutsk regions, and Novosibirsk city, and parameters of the most effective model of approximation

Settlement Samburg, Tyumen region				
			G, P	
Age	N0	N1	mG/G	mP/P
50–80	24	7	0.2/1.7	**4.2**/15.4
39–49	36	4	5.0/19.3	**3.5**/11.1
18–38	71	7	3.2/18.5	**4.2**/44.7
<18	29	–	–	**3.3**/28.1
Settlement Maloe Goloustnoe, Irkutsk region				
			G, P	
Age	N0	N2	mG/G	mP/P
60–80	10	3	3.0/9.9	–
40–60	32	4	5.8/7.3	**4.2**/24.6
18–40	23	1	–	**4.0**/23.7
<18	15	–	–	**3.1**/15.3
Novosibirsk city				
			G, P	
Age	N0	N3	mG/G	mP/P
–	40	40	–	1.4

The standard error of the sample means d(mP) = 10–15 %, d(mG) = 30–40 %. The control group consisted of persons from Novosibirsk, the number of analyzed lymphocytes exceeding 100 in each blood sample. *N0*, total number of examined persons; *N1*, number of persons in whose blood samples the lymphocytes were not activated; *N2*, number of persons in whose blood samples the number of activated lymphocytes did not exceed 30; *N3*, the number of persons in whose blood samples the number of analyzed lymphocytes exceeded 100

abnormalities is decreasing in the children's group in comparison with the adult group. This is not in conflict with our conclusion that aging increases the probability of the bystander processes, and youth and resistance in persons play the main role in selection.

We would like to hope that the instability processes relax with subsequent generations. But the experiments performed on yeasts have shown that irradiation induces chromosomal instability which continues in hundreds of generations (Korogodin and Bliznik 1972). Perhaps humans have strong compensatory mechanisms which decrease the instabilities.

6.3.6 Risk of Chromosomal Instability for Individuals

The bystander processes result in the accumulation of cells with abnormalities and an increase of their frequency in blood. The Poisson parameters determine the

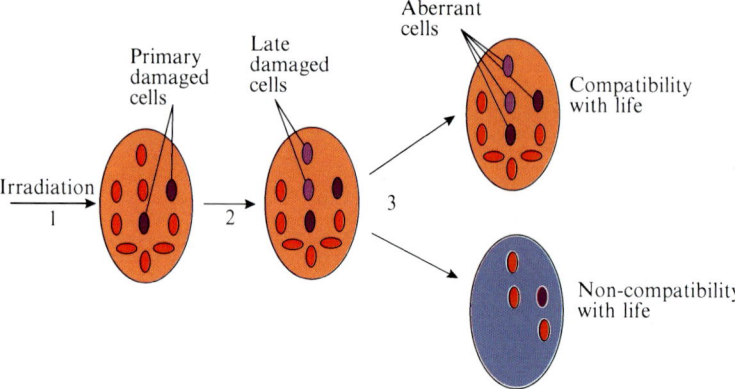

Fig. 6.13 Scheme of the appearance of blood lymphocyte cells with abnormalities induced by radiation stress. The stress factor leads to the increased number of cells with abnormalities in radioresistent organisms and to the decreased number of PCs in the radiosensitive one. The case of the increased number of abnormal cells of lymphocytes can lead to compensation processes in the resistant organism. On the contrary, the sensitive individual can die due to lymphocytes pool depletion and blood diseases

risk of chromosomal instability in the resistant group of persons. For example, the frequency of abnormal lymphocyte cells is 3.2 ± 0.3 for 50 % of "young" samples from the Tyumen and Irkutsk regions, where it is 5 % of the control group (Table 6.7). The geometric subpopulation is related to the sensitive individual fraction, where instability leads to lymphocyte depletion and death of the individuals. The geometric statistics account for the individuals with lymphocyte deficiencies and blood diseases. These persons form "the group of risk" which can be determined by the geometric value. The parameters of the Poisson and geometric distributions have demonstrated that the risk of radiation-induced chromosomal instability and selection prevails in the "old" generation.

6.3.7 Mechanisms of Genetic Adaptation for Humans

The hypothesis of radiation adaptation can be formulated for the human population on the basis of the discussed models.

Radiation stress induces: (1) primary injuries of cells, (2) the late processes which lead to the damaging of other cells, and (3) selection, at which some of the individuals with increased frequency of lymphocyte abnormalities survive, and the others die due to depletion of the lymphocyte pool. These processes are shown in the scheme (Fig. 6.13).

The adaptation hypothesis in mathematical terms (Florko and Korogodina 2007) can be formulated as follows:

6.3 Statistical Modelling for Persons Living in the Radiation – Polluted Areas

- Statistics of the primary damaged cells are described by the Poisson law, as are statistics of rare and independent events by

$$P_n = \frac{\alpha_0^n}{n!} e^{-\alpha_0},$$

where P_0 – is the probability of the appearance of n cells with abnormalities, and α_0 is the parameter of the Poisson distribution of primary damages;
- The secondary intercellular processes can be initiated in some subpopulations of blood cells that lead to the appearance of new cells with abnormalities. It results in the changing of the Poisson parameter a which depends linearly on time $a = \lambda t + a_0$,

$$P_n(a) = \frac{\alpha^n}{n!} e^{-\alpha} = \frac{(\lambda t + \alpha_0)^n}{n!} e^{-(\lambda t + \alpha_0)}.$$

- The appearance of new abnormal cells increases the material for selection. The environmental conditions determine the intensity of the appearance of secondary abnormalities. The compensatory-recovery processes are induced in the resistant organisms, whereas the sensitive ones die. Thus, the resistance and compensatory-recovery processes lead to "the adaptation." The probability to survive at time t can be described by the formula (Feller 1957)

$$G(t) = 1 - \sum_{i=1} \alpha_i e^{-\mu_i t},$$

where $\alpha_i > 0$ and $\mu_i > 0$ – are a set of parameters dependent on the intensity of the appearance of cells with abnormalities and characteristics of the compensatory-recovery processes, respectively.

In this case, the distribution of individuals on the number of abnormal cells can be described by the combination of the geometric and Poisson $(G + P)$ distributions. The distribution of individuals on the number of cells with abnormalities in blood samples preserves the Poisson type in the case of high resistance of the organism (strong compensatory-recovery processes) and the selection absence. So, the increased number of abnormalities should be observed.

This motivation is related to the analysis of distributions of individuals on the number of cells with abnormalities in blood lymphocytes. Indeed, we should consider distributions on the frequency of abnormal cells among the proliferated ones. In this case, the distribution of individuals on the number of PCs is important (Florko et al. 2009).

The distribution structure of the individuals on the frequency of cells with abnormalities allows one to analyze the law of appearance of cells with abnormalities: the spontaneous or late process of cells with abnormalities appearance (P); adaptive process on the basis of the late appearance of cells with abnormalities coupled with selection (G).

6.4 The Appearance of Multi-aberrant Cells Induced by the Radiochemical Industry

It was published that multi-aberrant cells were observed in the blood of individuals who worked in the Siberian radiochemical industry in the settlement of Seversk (Bochkov et al. 1972). Let us consider the appearance of multi-aberrant cells using the adaptation model by this example.

A.N. Chebotarev has offered a mathematical model which describes these data (Chebotarev 2000). He used model "P + G", and the efficiency of the modelling was estimated by the χ^2-criterion (Table 6.8). According to this model, the cell population is divided into two subpopulations. In the first subpopulation, the chromosomal aberrations appear independently under external factors and are distributed according to the Poisson law; in the second, there is a factor which promotes the appearance of aberrations under the same external factor. The author analyzed the model parameters to investigate the characteristics of low-radiation effects. For all groups of persons, the sample means of geometric distribution are high, although their relative values are small. It is surprising that the G-sample mean is so high for the control group. The main quota of the distributions belongs to the Poisson component whose sample mean is negligible.

This model can be added with the permanent adaptation process in intact cells described by the geometric law (see Sect. 3.4.2). Table 6.9 presents the results of the approximations performed according to this model (Korogodina et al. 2010a). It is expected that additional information can be received.

The data have demonstrated that radiation effects are the same for the "inhabitants" and "workers irradiated with 0.18÷0.37 Sv" because the characteristics for the resistant as well as for the sensitive groups are identical. The average number

Table 6.8 Parameters of the model P + G offered by A.N. Chebotarev (2000)

Group	P value	G value	P sample mean	G sample mean	G relative value, %
Inhabitants	4347.0	116.0	0.0246	1.435	3.68
Workers, 0.18÷0.37 Sv	6067.4	347.6	0.339	1.008	5.42
Workers, 0.93÷1.57 Sv	5570.5	274.5	0.289	1.091	4.70

Table 6.9 Statistical modelling on blood lymphocytes of people who lived in the town of Seversk and workers in its radiochemical industry

Group	Modelling	G1	mG1	G2	mG2	P	mP
Inhabitants	G1 + G2	0.96	0.03	0.04	1.17	–	–
Workers, 0.18÷0.37 Sv	G1 + G2	0.92	0.03	0.07	0.82	–	–
Workers, 0.93÷1.57 Sv	G1 + P	0.03	1.49	–	–	0.97	0.04

Standard errors of parameters did not exceed 10–15 %. The data on chromosomal aberrations are published in (Bochkov et al. 1972). The hypotheses "P" and "G" do not satisfy the criterion χ^2 (p > 0.05)

of CAs per cell is approximately 1 for 4 and 7 % of individuals from these two groups, respectively. Such percentage is very high. For 3 % of individuals from the third group of persons, the average number of aberrations per cell is about 1.5, and 97 % of cells have primary damages with an average number per cell of about 0.04. This model has shown identical genetic processes in the first and second groups which correspond to the inhabitants and managerial workers; serious chromosomal instability is observed in the blood of both groups. Irradiation of the third group (the technical workers) caused direct DNA damages (the Poisson component) for a significant number of individuals, and high chromosomal instability in the others.

6.5 Summary

In this chapter, the radiation effects were analyzed for the populations living in three sites which experienced nuclear test fallout in the middle of the previous century. These are the Tundra Nenets from the Yamalo Nenets autonomous area and inhabitants of the settlements of Maloe Goloustnoe and Listvyanka in the Pribaikal'e region polluted due to the Novaya Zemlya and Semipalatinsk tests. The statistical modelling was performed for four generations of their inhabitants: "parents" who had immediately experienced the radiation impacts, their "children," "grandchildren," and "great-grandchildren" younger than 18 years. The distributions of individuals were analyzed on the occurrence frequency of abnormal lymphocyte cells among the proliferated ones. The analysis has shown different consequences for these populations.

What determines the G- or P-type of individuals' distribution? The geometric and Poisson distributions of the individuals have been revealed. The geometric component means the selection of individuals caused by the instability process in the blood of persons. This statement was checked: it is shown that the geometric component is formed from individuals with bad activated lymphocyte cells and lymphocyte pool depletion. The Poisson component shows accumulation of abnormal cells. It is revealed in younger generations and can be interpreted as "effect of youth" because aging contributes to the bystander effect and relaxation. It seems that the first assumption should be accounted for because the experiments on irradiated yeasts have demonstrated the lasting chromosomal instability throughout hundreds of generations (Korogodin et al. 1977).

Consequences of radiation fallout for the Northern population. It is concluded that the worst situation is observed in the Northern population. The studies have demonstrated a series of "G + P" distributions for the "parents," their "children," and "grandchildren"; the "great-grandchildren" group is the Poisson-distributed. It suggests the instability process is accompanied by blood diseases and the death of individuals in elder groups, and elevated frequency of lymphocytes with abnormalities in the young group two times higher than in the control group. The selection process continues across three generations, and it suggests prolonged irradiation or strong fallout impact. Radioactivity is accumulated in the lichens and mosses which are food for the reindeer in the Polar tundra. The ^{137}Cs activity in venison is high enough to be a constant source of chronic internal radionuclide irradiation for the Tundra Nenets. The second suggestion needs additional studying.

Consequences of radiation fallout for persons living in the Pribaikal'e sites. The radiation influence is not so strong for the Pribaikal'e region. The same model "G + P" has been revealed only for the "parents" and "children" groups from Maloe Goloustnoe that suggests a lower irradiation in comparison with the Northern area and the natural radiation purification of the polluted territory. The "grandchildren" and "great-grandchildren" groups are Poisson-distributed and their P-sample means are elevated in comparison with the control group. The Listvyanka sample ("parents") is P-distributed, and we can assume that the radiation impact was less in the Listvyanka site than in Maloe Goloustnoe.

Influence of the radiochemical industry in the town of Seversk on chromosomal instability in the blood of the workers and town inhabitants. The statistical modelling has been performed to study how the radiochemical industry influences the occurrence of multi-aberrant cells in the blood of its workers and inhabitants of the town. The elevated chromosomal instability has been found in the blood of the managerial staff and inhabitants (it is 1 damage per cell), which was described by the combination of two geometric components in both cases. The second component is formed from the sensitive lymphocyte cells. The technical workers (97 %) experience strong irradiation which leads to direct DNA damages, and elevated chromosomal instability (1.5 damages per cell) is expected for others (3 %).

References

Abil'dinova GZ, Kuleshov NP, Sviatova GS (2003) Chromosomal instability parameters in the population affected by nuclear explosions at the Semipalatinsk nuclear test site. Genetika 39:1123–1127 (Russian)

Antonova E, Osipova LP, Florko BV et al (2008a) The comparison of distributions of individuals with normally and poorly stimulated blood cell activity on the frequency of aberrant cells' occurrence in blood lymphocytes. Rep Russ Mil-Med Acad 1(3):73 (Russian)

Antonova E, Osipova LP, Sen'kova NA et al (2008b) The comparison of distributions of individuals under 18 and older on the frequency of aberrant cells' occurrence in blood lymphocytes in the samples of individuals living in the settlements of Maloe Goloustnoe and Listvyanka in Irkutsk region. Rep Russ Mil-Med Acad 1(3):90 (Russian)

Averbeck D (2010) Non-targeted effects as a paradigm breaking evidence. Mutat Res 687:7–12

Bezdrobna L, Tsyaganok T, Romanova O et al (2002) Chromosomal aberrations in blood lymphocytes of the residents of 30-km Chernobyl NPP exclusion zone. In: Imanaka T (ed) Recent research activities about the Chernobyl NPP accident in Belarus, Ukraine and Russia. Kyoto Universtiy Research Reactor Institute (KURRI-KR-79), pp 277–287

Bochkov NP, Yakovenko KN, Chebotarev AN et al (1972) Distribution of the damaged chromosomes on human cells under chemical mutagens effects in vitro and in vivo. Genetika 8:160–167 (Russian)

Bochkov NP, Chebotarev AN, Katosova LD et al (2001) The database for analysis of quantitative characteristics of chromosome aberration frequencies in the culture of human peripheral blood lymphocytes. Genetika 37:549–557 (Russian)

Bochkov NP (1993) Analytic review of the cytogenetic investigations after the Chernobyl accident. Bull Russ Acad Med Sci 6:51–56 (Russian)

Bolegenova NK, Bekmanov BO, Djansugurova LB et al (2009) Genetic polymorphisms and expression of minisatellite mutations in a 3-generation population around the Semipalatinsk nuclear explosion test-site, Kazakhstan. Int J Hyg Environ Health 212:654–660

Boltneva LI, Izrael YA, Ivanov VA et al (1977) Global ^{137}Cs and ^{90}Sr pollution and doses of the external radiation exposure in the USSR territory. Atomnaya Energiya 42:355–360 (Russian)

Cao S, Deng Z, Zhen Z et al (1981) Lymphocyte chromosome aberrations in personnel occupationally exposed to low levels of radiation. Health Phys 41:586

Chebotarev AN (2000) A mathematical model of origin of multi-aberrant cell during spontaneous mutagenesis. Rep RAS 371:207–209 (Russian)

Daillant O, Boilley D, Gerzabek M et al (2004) Metabolised tritium and radiocarbon in lichens and their use as biomonitors. J Atmospheric Chem 49:329–341

Davis W Jr, Ronai Z, Tew KD (2001) Cellular thiols and reactive oxygen species in drug-induced apoptosis. J Pharmacol Exp Theor 296:1–6

Dubrova YE, Jeffreys A, Nesterov VN et al (1996) Human minisatellite mutation rate after the Chernobyl accident. Nature 380:683–686

Dubrova YE (2003) Radiation-induced transgenerational instability. Oncogene 22:7087–7093

Eliseeva IM, Iofa EL, Stoian EF et al (1994) An analysis of chromosome aberrations and SCE in children from radiation-contaminated regions of Ukraine. Radiats Biol Radioecol 34:163–171 (Russian)

Feller W (1957) An introduction to probability theory and its applications. Wiley/Chapman & Hall, Limited, New York/London

Florko BV, Korogodina VL (2007) Analysis of the distribution structure as exemplified by one cytogenetic problem. PEPAN Lett 4:331–338

Florko BV, Osipova LP, Korogodina VL (2009) On some features of forming and analysis of distributions of individuals on the number and frequency of aberrant cells among blood lymphocytes. Math Biol Bioinform 4:52–65

Korogodin VI, Bliznik M (1972) Formation of radioraces by yeasts. Comm. 1. Radioraces of diploid yeasts *Saccaromyces ellipsoideus vini*. Radiologiya 12:163–170 (Russian)

Korogodin VI, Bliznik KM, Kapultsevich YG (1977) Regularities of radioraces formation in yeasts. Comm. 11. Facts and hypotheses. Radiologiya 17:492–499 (Russian)

Korogodina VL, Florko BV, Osipova LP et al (2010a) The adaptation processes and risks of chromosomal instability in populations. Biosphere 2:178–185 (Russian)

Korogodina VL, Florko BV, Osipova LP (2010b) Adaptation and radiation-induced chromosomal instability studied by statistical modeling. Open Evol J 4:12–22

Lazjuk G, Satow Y, Nikolaev D et al. (1999) Genetic consequences of the Chernobyl accident for Belarus Republic. In: Imanaka T (ed) Recent research activities about the Chernobyl NPP accident in Belarus, Ukraine and Russia. Kyoto University Research Reactor Institute (KURRI-KR-7), pp 174–177

Liden K (1961) ^{137}Cs burdens in Swedish Laplanders and reindeer. Acta Radiol 56:64–65

Matveeva VG, Sablina OV, Eremina VR et al (1993) Cytogenetics of the inherent pathology in the inhabitants of the Altai radiation-polluted zones. In: Genetic effects of the anthropogenic environment factors. Novosibirsk 1: 5–7 (Russian)

Medvedev VI, Korshunov LG, Chernyago BP (2005) Radiation effect of Semipalatinsk nuclear polygon on the South Siberia (investigations of several years on the East and Middle Siberia and the comparison with the data on the West Siberia). Siberia Ecol J XII:1055–1071 (Russian)

Medvedev VI, Korshunov LG, Chernyago BP et al (2009) Radiation effect of the Semipalatinsk nuclear polygon on the South Siberia. Probl Biogeochem Geochem Ecol 2:57–65 (Russian)

Mikhalevich LS (1999) Monitoring of cytogenetic damages in peripheral lymphocytes of children living in radiocontaminated areas of Belarus In: Imanaka T (ed) Recent research activities about the Chernobyl NPP accident in Belarus, Ukraine and Russia. Kyoto University Research Reactor Institute (KURRI-KR-79), pp 178–188

Moorhead PS, Howell PC, Mellman WJ (1960) Chromosome preparations of leucocytes cultured from human peripheral blood. Exp Cell Res 20:613–616

Nepomnyaschih AI, Chernyago BP, Medvedev VI et al (1999) Radioecological condition of the Irkutsk region territory. Joint report on the program "Radon" (1998) of the Institute of geochemistry SB RAS, Irkutsk State University, St. Petersburg Institute of Radiation Hygiene, Irkutsk, pp 48–49

Nizhnikov AI, Nevstrueva MA, Ramzaev PV et al (1969) Cs137 in food chain lichen-reindeer – man in Far North USSR (1962–1968). Atomizdat, Moscow

Osipova LP, Posukh OL, Koutzenogii KP et al (1999) Epidemiological studies for the assessment of risks from environmental radiation on Tundra Nenets population. In: Baum Stark-Khan C et al (eds) NATO science series (2): environmental security, vol 55. Kluwer, Dordrecht, pp 35–42

Osipova LP, Posukh OL, Ponomareva AV et al (2000a) Medicogenetics investigations of the population of Tundra Nenets and evaluation of radiation situation in the region of their habitation. Siberian Ecologycal J 1:61–65 (Russian)

Osipova LP, Ponomareva AV, Shcherbov BL et al (2000b) Aftermath of radiation effect in population of Tundra Nenets inhabitants, Purov region, YaNAD. In: Proceedings of the international conference on "Modern Problems of Radiobiology, Radioecology and Evolution". JINR, Dubna, 2000, pp 200–212 (Russian)

Osipova LP, Koutsenogii KP, Shcherbov BL et al (2002) Environmental radioactivity for risk assessment of health status in nature and human population of Northern Siberia. In: Proceedings of the 5th international conference on environmental radioactivity in the Arctic and Antarctic. St. Petersburg, 2002, pp 124–127 (Russian)

Osipova LP, Sen'kova NA, Gainer TA et al (2008) Cytogenetic assessment of the late consequences of the radiation factors on the inhabitants of the settlement Maloe Gouloustnoe in the Irkutsk region. Rep Russ Mil-Med Acad 1(3):134 (Russian)

Pilinskaya MA, Shemetun AM, Eremeeva MN et al (1991) The cytogenetic effect in the peripheral blood lymphocytes of persons with a history of acute radiation sickness as a result of the accident at the Chernobyl Atomic Electric Power Station. Tsitol Genet 25(4):17–21 (Russian)

Pinzino C, Capocchi A, Galleschi L et al (1999) Aging, free radicals, and antioxidants in wheat seeds. J Agric Food Chem 47:1333–1339

Scherbov BL, Malikova IN, Osipova LP et al (2006) Radioecology conditions in the territories of the Siberia native inhabitants at the turn of the XX-XXI centuries. Probl Biogeochem Geochem Ecol 3(3):520–531 (Russian)

Seabright MA (1971) Rapid banding technique for human chromosomes. Lancet 2:971–972

Sevan'kaev AV, Potetnia OI, Zhloba AA et al (1995) The results of the cytogenetic examination of children and adolescents living in radionuclide-contaminated regions of Kaluga province. Radiats Biol Radioecol 35:581–588 (Russian)

Shevchenko VA, Snigireva GP, Suskov II et al (1995) The cytogenetic effects in the population of the Altai territory subjected to ionizing radiation exposure as a result of the nuclear explosions at the Semipalatinsk proving grounds. Radiats Biol Radioecol 35(5):588–596 (Russian)

Shevchenko VA (1997) Integral estimation of genetic effects of ionizing radiation. Radiats Biol Radioecol 37(4):569–576 (Russian)

Shevchenko VA, Snigiryova GP (1999) Cytogenetic effects of the action of ionizing radiations on human populations. In: Imanaka T (ed) Recent research activities about the Chernobyl NPP accident in Belarus, Ukraine and Russia. Kyoto University Research Reactor Institute (KURRI-KR-79), pp 203–216

Tanaka K, Iida S, Takeichi N et al (2006) Unstable-type chromosome aberrations in lymphocytes from individuals living near Semipalatinsk nuclear test site. J Radiat Res (Tokyo) 47(Suppl A):A159–A164

The Ministry of Nature of RF (1992) Criteria for the assessment of ecological conditions in the territories to test zone in extreme ecological situation. The Ministry of Nature of RF, Moscow

References

Vorobtsova IE, Vorob'eva MV, Bogomazova AN et al (1995) The cytogenetic examination of children in the Saint Petersburg region who suffered as a result of the accident at the Chernobyl Atomic Electric Power Station. The frequency of unstable chromosome aberrations in the peripheral blood lymphocytes. Radiats Biol Radioecol 35(5):630–635 (Russian)

Yablokov AV, Nesterenko VB, Nesterenko AV (2009) Chernobyl: consequences of the catastrophe for people and the environment. Ann N Y Acad Sci 1181:vii–xiii, 1–327

Chapter 7
Conclusion

Abstract Here, we will summarize our knowledge of adaptation and chromosomal instability in terms of inter- and intracellular processes. This point of view presents these phenomena as a natural law regulating survival of organisms in their ecological niches. We emphasize the role of instability in providing the non-linearity characterizing low-radiation effects. Statistical modelling is considered as instrumental to the study of low-dose effects. Four issues will be discussed below: connection of adaptation with instability and selection processes, general features of adaptation and instability, consequences of low-radiation fallout for nature and humans, and how adaptation and instability processes can be described statistically.

Keywords Inter- and intracellular adaptation processes • Instability • Selection • Transgenerational instability • Effects of radiation fallout • Synergism • Risks of adaptation processes • Statistical modelling

7.1 Adaptation, Genetic Instability, and Selection Processes

An aged Russian geneticist-agronomist[1] told one of the authors of this book that the breeding plants growing on the edge of a cropped field often had mutations. He explained this phenomenon by the influence of non-optimal conditions on the mutation rate and adaptation. The influence of environmental stress conditions on mutagenesis was known long ago.

Beginning of the low-dose-effect investigations. Radiation is a well-measured factor, which was used to investigate mutagenesis (Timofeeff-Ressovsky et al. 1935). Until the 1950s, scientists studied the effects using the middle-lethal dose irradiation related to post-radiation recovery: dose irradiation, number of cell

---

[1] Prof. Nikolaj F. Batygin in the 1980s.

V.L. Korogodina et al., *Radiation-Induced Processes of Adaptation: Research by statistical modelling*, DOI 10.1007/978-94-007-6630-3_7,
© Springer Science+Business Media Dordrecht 2013

hboxdamages, recovery or death of cells (Korogodin 1966). Such dependence is presented in Chap. 4 (Fig. 4.1a): irradiation at acute doses of 25–80 Gy related to the linear increase of both the frequency of cells with abnormalities in seedlings meristem and death of seedlings. The first study of the adaptation phenomenon was an investigation of the non-linearity induced by low-dose irradiation performed by N.V. Luchnik in the middle of the previous century (Luchnik 1958). Similar experiments on pea seeds were made at the JINR (see Fig. 4.1b) (Korogodina et al. 1998). The studies of low-dose effects showed induction of chromosome damages, dramatic decrease of survival and increased radioresistance (Korogodina et al. 1998). In the 1990s, many scientists described the effects induced by low irradiation (reviewed in (Averbeck 2010)).

The bystander effect. Investigations of the bystander effect explained the non-linearity: radiation stress induces the "non-targeted" process of cells' damaging; injuring of one cell leads to the appearance of some damages in the neighboring cells (Mothersill and Seymour 2001). So, the bystander effect initiates genomic instability. In its turn, genomic instability leads to selection that assumes the adaptation process. Transition from the low-dose and dose-rate irradiation to the middle-lethal equivalent switches cells from the adaptation strategy (when the damages accumulate until balanced with the environment) to the recovery of their damages.

Statistical modelling of the appearance of cells with abnormalities. Statistical modelling of the appearance of cells with abnormalities demonstrates a connection of adaptation to the instability and selection processes. Statistical modelling for pea seeds presents the distribution of seedlings on the number of abnormalities as the combination of Poisson and geometric distributions (see Sect. 3.4.1), which are the resistant and sensitive components. The Poisson law reflects rare independent injuring which can be radiation-induced "bystander" damages accumulated in the resistant subpopulation. The geometric law describes the cells damaging processes until stopped due to "success": balance between the new set of cells and environmental conditions, or death as the result of too many damages. The selection leads to dramatic mortality of seeds in the more damaged sensitive subpopulation. Both P- and G-components describe the instability process, but it is accompanied by accumulation in a resistant subpopulation, and by selection in a sensitive one. We can calculate the contributions of instability coupled with accumulation, selection, and mortality.

Let us consider low-dose irradiation of pea seeds at 7 cGy with 0.3 cGy/h which decreases both seed survival and frequency of cells with abnormalities in meristem of seedlings (see Fig. 4.1). The leading role of the G-regularities in this case is clear: the sample mean is high, and the relative value is the lowest (see Table 4.4). No wonder that seed survival strongly decreases.

Figure 5.6 (Chap. 5) reflects the non-linear dependence of P- and G-subpopulations on the dose-rate radiation in ecology: both G- and P-values decrease along the NPP line and reach the minimum at the border of the sanitary zone, and one can assume the strongest selection process there. The contribution of P-subpopulation increases in the sanitary zone (see Fig. 5.6), which means an increase in seed survival (see Fig. 5.3). Besides, the dramatic fall of the G-sample mean is observed, which assumes dominance of the primary damages in the sanitary zone. So, the instability and selection processes are observed around the NPP, but primary irradiation contributes significantly in the sanitary zone.

Statistical modelling of the appearance of chromosomal abnormalities. The low-intensive processes of chromosomal instability are constant in cells (see Chap. 4). The additional environmental factor induces a chromosomal instability in the sensitive subpopulation of cells, which can be described by the geometric law with the other parameters. So, the appearance of multiple aberrations in cells can be presented by the combination of two geometric distributions of cells on the number of chromosomal abnormalities (see Sect. 3.4.2). The second geometric distribution can be replaced by the Poisson law in the case of increased radiation intensity. One can see that direct DNA damages prevail at lab irradiation of pea seeds with a dose rate of 19.1 cGy/h (see Table 4.8) and near the radiation sources (see Table 5.5). The same regularities are observed in the radiochemical industry for technical workers (see Table 6.9).

Statistical modelling of the appearance of proliferated cells. Adaptation affects the proliferation activity of cells. Some cells change their proliferative activity, others die, and the resting cells can be stimulated to divide (see Sect. 4.1.3). Thus, the proliferative cells can appear in three independent subpopulations: two subpopulations reflect the heterogeneous character of the proliferated pool and the third subpopulation corresponds to the resting cells stimulated to proliferation (Luchnik 1958). The appearance of the proliferated cells can be described by the lognormal law (see Sect. 3.4.3). The lognormal law can be observed at negligible environmental factors, whereas increasing of one of them (for example, radiation) leads to mutagenesis and selection which is described by the geometric statistics. Similar regularities are observed near the Balakovo NPP and the JINR facility (see Fig. 5.5).

Radiation stress induces adaptation mechanisms which regulate the chromosomal instability process and the number of proliferated cells. The mortality of seeds increases due to the elimination of cells and the failure of seeds to germinate in the sensitive (geometric) subpopulation of seedlings. We would like to speculate that a stress factor induces the bystander mechanism coupled with selection (filtration) to adapt a population in its ecological niche. Adaptation includes increasing variability (quickly elevating the G-sample mean) and dramatic decreasing of the number of germinated seeds. Here, we remember N.W. Timofeeff-Ressovsky (1939) who emphasized two requirements of evolution: a sharp increase in the evolutionary material coupled with partial death of a population because a new genotype would get an advantage only in this case.

7.2 Regularities of the Adaptation Processes

The adaptation process includes three components: primary and late damaging, and selection. These processes are determined by the intensities of the primary irradiation and late processes, and organism sensitivity. Let us consider the influence of these characteristics on the adaptation process.

Influence of dose-rate irradiation: Radiation intensity determines the type of genetic processes. We use statistical modelling to explain this thesis and substitute it for the following: "the type of the distribution depends on the radiation intensity."

Mathematical analysis has shown that a G-component appears when the average number of primary damaged cells is small (see Sect. 3.4.1). The resistant P-cells pass gradually with the dose-rate irradiation into the sensitive G-subpopulation until radiation intensity prevails in the late processes. Thus, the distribution of cells on the number of abnormalities obtains the Poisson character with an increase in the number of primary damages dependent on radiation intensity (see Sect. 3.4.2). These regularities were discussed above.

The distribution of proliferated cells is described by the lognormal law (see Sect. 3.4.3) (Florko and Korogodina 2007). Usually, the distribution of proliferated cells consists of two lognormal ones due to heterogeneity of cells. The low-dose irradiation leads to the appearance of the third lognormal distribution, which corresponds to the resting cells stimulated to divide. The other effect is that the increased radiation stress factor intensifies the selection, and the lognormal statistics of proliferation is replaced by the geometric law of selection. These regularities are demonstrated in Fig. 5.5.

Dependence of adaptation process on sensitivity: Selection is an instrument of adaptation: selection increases with the stress factor, and decreases immediately when the intensity of primary damages exceeds the late ones. The parameter of the G-distribution depends on the relation of the "communication" (signaling which regulates late damaging) and selection intensities. The G-sample mean obtains minor values with increasing repair intensity, but the character of the cells' distribution on the number of damages doesn't change (see Sect. 3.4.2).

The antioxidant status determines the protection systems of an organism, and here the dependence of adaptation on these characteristics is studied. Ecological investigations observed the dependence of relative values of seed- and cell distribution on antioxidant status (see Figs. 5.6 and 5.7b). This verifies that antioxidants are radioprotectors and they influence selection. The antioxidant status also influences the sample means of seed distributions (see Fig. 5.6a). No influence of antioxidant status was revealed on the cell distribution sample means (see Fig. 5.7a), perhaps because the sample sizes were not sufficient. The other hypothesis is low NPP fallouts: by our calculations each seedling's meristem experiences the influence of even one γ-quantum per three months in the 20-km NPP zone (Korogodina and Florko 2007). We suppose that multiple DNA damages could appear due to the affected mass cellular structures. This conclusion is in agreement with the opinion concerning the role of epigenetic factors in low-dose effects (Kovalchuk and Baulch 2008).

Aging decreases resistance of cells and organisms, and lab experiments have demonstrated the important role of resistance in reactions of old pea seeds (see Chap. 4). In old seeds, the non-viability is related to frequency of cells with abnormalities, and the partial correlations analysis has shown the close connection of frequency of cells with abnormalities as well as seed survival with the geometric subpopulation (see Table 4.5). The geometric subpopulation is considered to play an important role in aging seeds, because the non-surviving fraction increases primarily as a result of diminution of "geometric" seeds, and the frequency of abnormal cells also increases with the sample mean of the geometric distribution.

The important role of resistance is demonstrated in investigations of human populations that experienced a nuclear test fallout. The geometric component was eliminated from the individual distribution on the frequency of cells with abnormalities in blood lymphocytes of younger generations (see Fig. 6.11).

Dependence of adaptation process on the intensity of late processes: Dependence of adaptation on the intensity of late processes is not considered here in detail. The sample means are completely determined by the relation of late processes and repair rates (see Sects. 3.4.1 and 3.4.2), and reactive oxidative species are involved in both the bystander effect and genomic instability (Morgan 2003a, b; Mothersill et al. 2000). Besides, the antioxidant status characterizes the resistance of cells and organisms.

7.3 Instabilities Induced by Low-Radiation Fallout

Now it is time to discuss the effects caused by radiation impacts which do not exceed the natural background. These are territories polluted by fallout of the nuclear power plants (Balakovo NPP) and the nuclear tests carried out 50 years ago (Pribaikal'e region, North Siberia). The processes of chromosomal instability are observed near the Balakovo NPP, and continue across four generations of the human population living in the territories that experienced radiation fallout many years ago. The radiation effects can be aggravated by climate conditions (hot summer in the Volga region or permafrost in the Polar tundra), age of organisms and other factors.

Radiation effects in the 30-km zone of the nuclear power plant: Our aim was to show the radiation effect near the standard operating NPP the fallout of which does not exceed the background. For investigations, we chose the plantain populations growing in the 30-km zone of the Balakovo NPP (Saratov region) and three control groups in the Saratov region and in the JINR (Dubna, Moscow region) territory. The plantain seeds were collected in 1998 and 1999, and the summer of the latter year was hot. In 1998, some chemical-polluted sites were tested, but they were excluded in 1999.

In 1998, the radiation effect was not strong, and the correlation was observed between seed mortality and frequency of cells with abnormalities in seedling meristem. In 1999, the radiation effect was strong (see Figs. 5.6 and 5.7), and the correlation between mortality and frequency of abnormal cells disappeared. We could suspect the synergic effects induced by radiation and heat stresses because the influence of chemical pollution was excluded (see Sect. 8.5), and the same synergism of radiation and heat stresses was observed in the lab experiments (see Sect. 4.3.4). The seed survival fell to 20–30 % (see Table 5.3) in nature and 30 % in the lab (see Table 4.8), which is critical for the plant population (Preobrazhenskaya 1971). The combination of radiation fallout and a hot summer (1999) led to the instability and selection processes which were revealed due to decreased relative values of both G- and P-distributions and their increased sample means (see Table 5.5, Figs. 5.6 and 5.7). The contribution of P-subpopulation increases in the sanitary zone that increases seed survival.

It is necessary to emphasize the role of humidity at the tested site in determining the seed antioxidant status. Two of the tested populations grew not far from each other in the town of Balakovo, but in the wet and arid sites. Seed survival of the first population was low but the frequency of cells with abnormalities was not high either (see Table 5.4). The second population was characterized by the inverse values. This paradox can be explained by instability and selection processes: the P-law dominates in the resistant population, and the G-statistics prevail in the sensitive one (see Table 5.6).

The studies of the health of the Balakovo inhabitants (1998) demonstrated a higher frequency of hematogenesis diseases in comparison with the control town in the Saratov region (by 1.2–7.6 times) (Dodina 1998). The author assumes that the morbidity elevation is caused by chemical and low-dose irradiation factors. The mathematical examination of chemical (see Sect. 8.5) and radiation contributions into chromosomal instability showed radiation-induced adaptation processes coupled with instabilities and selection. Of course, the chemical industry contributes to the bad health of the Balakovo inhabitants, but the high correlation of chromosomal instability and dose-rate fallout has verified that NPP fallout is very dangerous, especially in the hot summer.

How many γ-quanta can induce bystander effects? Laboratory studies have shown that irradiation with 0.3 cGy/h (intensity $\sim$1 γ-quanta/cell nucleus/min) is the most effective for inducing bystander effects (Korogodina et al. 2005). On the other hand, the number of primary Poisson damages should be minimal at this dose rate. A probability of double hits/nucleus/min would increase with the dose rate, whereas the bystander effect decreases significantly. It means that one γ-quantum can induce the bystander effect.

In natural experiments (see Chap. 5), γ-rays are mostly a part of irradiation near the NPP; their intensity is $\sim$1 γ-quanta/nucleus/3 months in immediate proximity to the NPP (sanitary zone). This means that each nucleus could be damaged in the vegetation period. In plant populations growing at the border of the sanitary zone, the number of NPP fallout γ-quanta is 10^{-3} times lower. The bystander effect was registered in all these populations. This does not seem too incredible because it has been shown that the irradiation comparable with the background induces changes in cellular membranes (Mothersill et al. 2004).

Continued instability across generations: The sites which experienced nuclear test fallout many years ago were chosen to study transgenerational instability. These are the Polar tundra (see Figs. 6.1 and 6.4) and Pribaikal'e (see Figs. 6.3 and 6.6) sites. The frequency of cells with abnormalities is high in the individuals living there, although the nuclear explosions were in the 1950s.

Statistical modelling has shown an increased P-sample means for the youngest groups of individuals living in the settlements of Samburg and Maloe Goloustnoe (see Table 6.7; Fig. 6.12) in comparison with the control that assumes chromosomal instability in the fourth generation. Chromosomal instability leads to lymphocyte pool depletion, blood diseases, and increased mortality. Resistance (or age) plays the main role in human mortality: samples of older individuals are characterized by the geometric distribution on the number of lymphocytes with chromosomal abnormalities (see Fig. 6.12) and by the lymphocyte depletion (see Table 6.7) in these sites. The youngest samples are P-distributed, which is the reason for low frequency of cells with abnormalities in blood lymphocytes in comparison with the older samples of individuals. We can assume the "G"-processes when the young Samburg group reaches 40–50 years old because the rings and dicentrics – the indicators of radiation instability – are observed in adult and children's samples (see Table 6.3).

It has been found that the intensity of bystander processes and selection is higher in Samburg because the G-component was revealed there in three generations of inhabitants (it was found in two generations in the Maloe Goloustnoe population). This strong radiation effect can be referred to as long irradiation due to permafrost and the northern food chain of "lichen-reindeer-man". All of the above should be considered as components of the prolonged adaptation process.

Calculation of risks of instability in ecology and epidemiology: The bystander effect leads to accumulation of cells with abnormalities which are accompanied by selection in the sensitive subpopulation of cells. Risks of instability related to the bystander effect can be calculated by means of parameters of P- and G-distributions.

The low P-sample mean indicates primary damages of cells, and its increased value shows the late processes (see Sect. 3.4.1). The risk of instability associated with selection can be determined by the G-distribution value.

7.4 How the Adaptation Process Can Be Presented Statistically

R.A. Fisher (1930) was the first to offer a fitness geometric model. H.A. Orr (2006) showed the universal character of distributions with tails in evolutionary models and the complementation of this phenomenon with selection (see Sect. 3.1).

The radiation-induced adaptation can be presented as a time-dependent process of genetic changes which lead to selection. Statistically this process can be described by the combination of the Poisson (or binomial) and geometric laws. The stress factor induces instabilities, and the appeared abnormalities are accumulated. This process goes rapidly in the sensitive fraction until we obtain "success": adaptation. The sensitive cell or organism can die during the adaptation process due to many abnormalities. The higher the resistance, the less selection during the adaptation process (see Figs. 5.6 and 5.7).

Let us consider the same ideas relative to the species abundance distributions. At the optimal environmental conditions the species abundance reaches its maximum, and its distribution can be described with the lognormal law as was offered by F. Preston (1962).[2] The species propagation can be compared as well as the proliferation of cells with the fragmentation process described by A.N. Kolmogorov (1986) with the lognormal law. Under the influence of the stress factor, the lognormal law should be replaced by the geometric one. I. Motomura (1932) suggested a hypothesis of priority ecological niche's occupation in terms of resource limitation that remains a pressure factor. The transformation of the "lognormal" ecosystems into the "geometric" ones was realized in the territory which had experienced radiation fallout of the Totsky explosion (Vasiliev et al. 1997).

At present, statistical modelling has been sufficiently developed to be an appropriate method to investigate the adaptation process and its components: instability, selection, and stimulation of proliferation (Korogodina and Florko 2007; Florko and Korogodina 2007; Florko et al. 2009). The method is based on distributions of population (organisms, cells) on the number of abnormalities and is qualified to investigate the processes induced by the low factors and accompanied by Darwinian selection in different systems. The parameters of distributions can be used to study the characteristics of adaptation processes and system resistance.

References

Averbeck D (2010) Non-targeted effects as a paradigm breaking evidence. Mutat Res 687:7–12
Dodina LG (1998) The health violation of individuals and adaptation mechanisms in the conditions of anthropogenic low-intensity factors. Dr. Sci. theses, St-Petersburg State Medical Academy named after II Mechnikov, St-Petersburg (Russian)
Fisher RA (1930) The genetical theory of natural selection. Oxford University Press, Oxford
Florko BV, Korogodina VL (2007) Analysis of the distribution structure as exemplified by one cytogenetic problem. PEPAN Lett 4:331–338
Florko BV, Osipova LP, Korogodina VL (2009) On some features of forming and analysis of distributions of individuals on the number and frequency of aberrant cells among blood lymphocytes. Math Biol Bioinform 4:52–65 (Russian)

---

[2]It evidently refers to the prosperous community (authors' note).

Kolmogorov AN (1986) About the log-normal distribution of particle sizes under fragmentation. In: The probabilities theory and mathematical statistics. Nauka, Moscow (Russian)

Korogodin VI (1966) The problems of postradiation recovery. Atomizdat, Moscow

Korogodina VL, Florko BV (2007) Evolution processes in populations of plantain, growing around the radiation sources: changes in plant genotypes resulting from bystander effects and chromosomal instability. In: Mothersill C, Seymour C, Mosse IB (eds) A challenge for the future. Springer, Dordrecht, pp 155–170

Korogodina VL, Panteleeva A, Ganicheva I et al (1998) Influence of dose rate gamma-irradiation on mitosis and adaptive response of pea seedlings' cells. Radiat Biol Radioecol 38:643–649 (Russian)

Korogodina VL, Florko BV, Korogodin VI (2005) Variability of seed plant populations under oxidizing radiation and heat stresses in laboratory experiments. IEEE Trans Nucl Sci 52:1076–1083

Kovalchuk O, Baulch JE (2008) Epigenetic changes and nontargeted radiation effects – is there a link? Environ Mol Mutagen 49:16–25

Luchnik NV (1958) Influence of low-dose irradiation on mitosis of pea. Bull MOIP Ural Department 1:37–49 (Russian)

Morgan WF (2003a) Non-targeted and delayed effects of exposure to ionizing radiation: I. Radiation-induced genomic instability and bystander effects *in vitro*. Radiat Res 159:567–580

Morgan WF (2003b) Non-targeted and delayed effects of exposure to ionizing radiation. II. Radiation-induced genomic instability and bystander effects *in vivo*, clastogenic factors and transgenerational effects. Radiat Res 159:581–596

Mothersill CE, Seymour CB (2001) Radiation-induced bystander effects: past history and future perspectives. Radiat Res 155:759–767

Mothersill C, Stamato TD, Perez ML et al (2000) Involvement of energy metabolism in the production of 'bystander effects' by radiation. Br J Cancer 82:1740–1746

Mothersill C, Seymour RJ, Seymour CB (2004) Bystander effects in repair-deficient cell lines. Radiat Res 161:256–263

Motomura I (1932) A statistical treatment of associations. Jpn J Zool 44:379–383 (Japanese)

Orr HA (2006) The distribution of fitness effects among beneficial mutations in Fisher's geometric model of adaptation. J Theor Biol 238:279–285

Preobrazhenskaya E (1971) Radioresistance of plant seeds. Atomizdat, Moscow (Russian)

Preston FW (1962) The canonical distribution of commonness and rarity: part I. Ecology 43:185–215

Timofeeff-Ressovsky NW (1939) Genetik und Evolution. Zeitschrift für inductive Abstammungs and Vererbungslehre 76:158–218 (German)

Timofeeff-Ressovsky NW, Zimmer KG, Delbrück M (1935) Über die Nature der Genmutation und der Genstruktur. Nachr Ges Wiss Gottingen FG VI Biol NF 1:189–245 (German)

Vasiliev AG, Boev VM, Gileva EA et al (1997) Ecogenetic analysis of late consequences of the Totskij nuclear explosion in Orenburg region in 1954 (facts, models, hypotheses). Ekaterinburg, Ekaterinburg

Chapter 8
Applications

Abstract This chapter presents the approximations, calculations and methods that require the use of statistical modelling. Special attention is devoted to the analysis of chemical pollution near a nuclear power plant which can induce cell damages or imitate the dependence of the distribution parameters on the average daily fallout. Eight topics will be discussed: Methods of approximation; Approximations of pea seeds distributions on the number of cells with abnormalities irradiated in the laboratory; Sample of the calculation of the synergic coefficient; Approximations of distributions for plantain seeds growing near a nuclear power plant; Analysis of the correlations of seed mortality $(1 - S)$ and parameters of the plantain seed distributions on the numbers of proliferated and abnormal cells with contamination of chemical pollutions in soil; Statistical modelling of the occurrence frequency of cells with abnormalities in blood lymphocytes of individuals.

Keywords Methods of approximation • Approximation efficiency • Synergic coefficient • Chemical pollution • Statistical modelling

8.1 Methods of Approximation

The methods of approximation are:

- ***The method of simple enumeration***: This method is used to search for the extremum of a function in bounded space depending on the argument's number $n \leq 3$. The range of the arguments variation should be covered by the uniform set, and we look for the extremum in the junctions of this set. If the number of junctions is large, the found extremum could be an approximation of the true extremum in the range.

  This method was used for an experimental data approximation with the step number ~ 50 for each argument and was realized in the preliminary calculations.

- **The random search method**: This method is used to search for the function extremum in the bounded region depending on the argument's number n > 3. The arguments values are randomly and uniformly chosen in the arguments variation range by means of a random-number generator. The argument set is chosen when the function reaches the extremum.
 The number of statistical samplings can be arbitrary. We used $N = 2 \times 10^6$. This method was realized by our own program.
- **The numeric method of the experimental data approximation by the χ^2 minimization**: This method is realized by the algorithm FUMILI (library CERNLIB of the FORTRAN, D510 programs (Sokolov and Silin 1962)). It was used for corrections of results provided by the first two methods.
- **The maximum-likelihood method**: This method was used for determination of the maximum plausibility function $l(\boldsymbol{\theta})$, which is necessary for the criteria AIC (Akaike 1974) and BIC (Schwarz 1978) calculation. The method was realized by algorithm FUMILI (Sokolov and Silin 1962).
- **The least-squares method**: The general idea of the least-squares method consists of sum minimization of the squared deviations of the model-predicted values from the observed ones. More precisely, assessment of the least-squares method for the θ parameters is obtained by minimization of the function R on θ, where

$$R = \frac{\sum_{i=1}^{N}(y_i - f(x_i, \theta))^2}{N - K},$$

N – is the amount of the experimental data, and K – is the number of model parameters. This method was used to analyze data and search for the model parameters, and realized by the standard program Origin.

The criteria for estimating the efficiency of the approximations: The following statistical criteria were chosen to estimate the regression: (i) R_{adj} is the determination coefficient corrected for degrees of freedom (Pytjev and Shishmarev 1983) (equivalent to the T-criterion known in radiobiology (Geraskin and Sarapultsev 1993)); (ii) the AIC criterion (the Akaike criterion imposes a limitation on the minimal information distance between the model and experimental distributions) (Akaike 1974); (iii) the BIC criterion (selection of the most probable models from the ensemble under condition of a priori equal probability of any of them) (Schwarz 1978). The most sensitive test for our tasks is the T-criterion (Geraskin and Sarapultsev 1993), which encourages good efficiency of approximation and fines for using large numbers of parameters.

The best hypothesis with respect to the majority of criteria was preferred. For close values of criteria, the simpler hypothesis was chosen. For a given number of experimental points, the number of model parameters is, as a rule, larger than the optimal one (Rakhlin et al. 2005); therefore, the stability of distributions was verified. The verification consisted in variation of the length of the partitioning interval upon construction of histograms. The interval length $(D_{max} - D_{min})/M$

8.1 Methods of Approximation

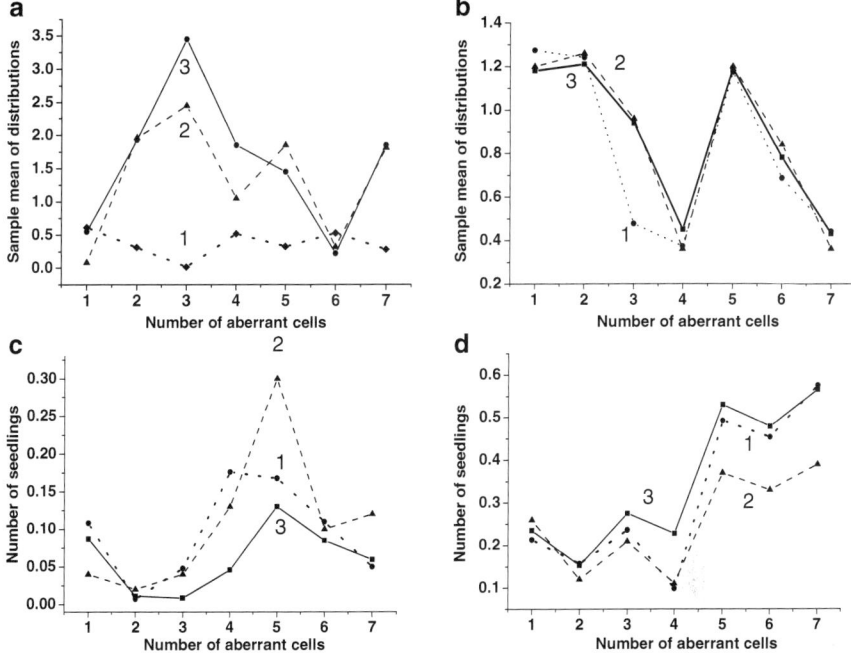

Fig. 8.1 Comparison of parameters of seed distributions with respect to the number of aberrant cells obtained using different programs: *1* – MATRIXER (optimization of three parameters), *2* – random search method and refinement using FUMILI (optimization of four parameters), and *3* – consecutive enumeration method (optimization of four parameters). Sample mean of the distribution "G" (**a**) and "P" one (**b**) number of rootlets in the distribution "G" (**c**) and "P" one (**d**) (Florko and Korogodina 2007)

was taken as the basis. Here, (D_{min}, D_{max}) is the interval of data variation, $M = [\log_2(N)] + 1$ is the number of partition intervals, and N is the number of data points. The verification showed the stability of the distributions.

The standard program, along with our own programs, was used for modelling. The program MATRIXER (Tsyplakov 2011) and consecutive enumeration methods were also used (Florko and Korogodina 2007). The search for optimal values of the model parameters was carried out in two stages. First, the initial approximation for the model parameters was found by the random search method (Turchak 1987) or the simulated annealing method (Kirkpatrick et al. 1983). Then, the parameter values were refined using regular procedures BFGS (Avriel 2003) or Newton (Turchak 1987). Three parameters were optimized. The comparison of parameters of seed distributions with respect to the number of aberrant cells obtained using different programs is presented in Fig. 8.1.

8.2 Approximations of Pea Seeds Distributions on the Number of Cells with Chromosomal Abnormalities Irradiated in the Laboratory

For approximation of the experimental data the following hypotheses were examined: single-component normal, binomial, lognormal, Poisson, and geometric distributions and their composition (Florko and Korogodina 2007; Korogodina and Florko 2007; Florko et al. 2009). For each distribution, different normalizing was used; the total samples of seedlings as well as the individual samples were tested; such examination did not reveal the principle differences in all cases. Assessment of the approximation efficiency was performed on several criteria.

The results of the experimental data approximations are presented, performed according to the hypotheses of the lognormal and sum of geometric and Poisson distributions. The standard program FUMILI and the χ^2 and T-criteria were used for approximations and assessment of their efficiency.

8.2.1 Experimental Data on the Pea Seeds Used for Approximations

Tables 8.1, 8.2, and 8.3 present experimental data for pea seeds on the number of cells with abnormalities in seedling root meristem of pea seeds. The numbers of seedlings (survived) is given in brackets and registered ana-telophases in apical meristem of rootlets, and the number of seedlings with number of abnormal cells corresponded to the intervals 0–2, 3–4, etc. are given.

8.2.2 Approximation Efficiency of Seed Distributions on the Number of Abnormal Cells

Both geometric and lognormal distributions have long tails. Table 8.4 presents the efficiency of the experimental data approximations by the lognormal and sum Poisson and geometric distributions.

The assessments of approximation efficiency indicated that the seeds of young groups are better described by the model "G + P". This model also has advantage for the old seeds irradiated with 19.1 cGy/h dose rate.

For the other groups of seeds, the efficiency of both models is comparable, except for the non-irradiated old seeds and the group of non-irradiated heat-stressed seeds. For these two non-irradiated groups, which are control for aging and heat factors, the lognormal model has advantage.

8.2 Approximations of Pea Seeds Distributions on the Number of Cells... 155

Table 8.1 Distribution of seeds on the number of abnormal cells in rootlet meristem of seedlings of young pea seeds

Dose rate, cGy/h	Number of seedlings	Number of ana-telophases	Number of seedlings with number of abnormal cells							
			0–2	3–4	5–6	7–8	9–10	11–12	13–14	<14
0	46(42)	2,202	11	9	10	6	3	1	0	2
0.3	42(35)	2,294	10	10	10	3	1	0	1	
1.2	45(42)	2,774	11	10	10	7	2	2		

Dose rate, cGy/h	Number of seedlings	Number of ana-telophases	Number of seedlings with number of abnormal cells					
			0–4	5–8	9–12	13–16	17–20	21–24
19.1	38(36)	2,144	23	9	1	1	1	1

Data histogramming was performed on the interval with equal length. The number of interval is $M = \left[\sqrt{N}\right] + 1$, where N – is the number of sample data

Table 8.2 Distribution of seeds on the number of abnormal cells in rootlet meristem of seedlings of old pea seeds

Dose rate, cGy/h	Number of seedlings	Number of ana-telophases	Number of seedlings with number of abnormal cells							
			0–2	3–4	5–6	7–8	9–10	11–12	13–14	14<
0	33(26)	2,202	10	8	3	4	1			
0.3	51(37)	2,294	6	8	8	7	3	3		2
1.2	53(42)	2,774	11	9	4	6	2	7	2	1
19.1	33(24)	2,144	13	8	2	1				

Data histogramming was performed on the interval with equal length. The number of interval is $M = \left[\sqrt{N}\right] + 1$, where N – is the number of sample data

Table 8.3 Distribution of seeds of "young + heat" group on the number of abnormal cells in rootlet meristem of seedlings of old pea seeds

Dose rate, cGy/h	Number of seedlings	Number of ana-telophases	Number of seedlings with number of abnormal cells				
			0–4	5–8	9–12	13–16	17–20
0	33(30)	2,202	4	13	8	5	0
0.3	39(26)	2,294	4	7	12	1	2

Dose rate, cGy/h	Number of seedlings	Number of ana-telophases	Number of seedlings with number of abnormal cells			
			0–6	7–12	13–18	19–24
0.3	43(16)	2,144	8	5	2	1

Dose rate, cGy/h	Number of seedlings	Number of ana-telophases	Number of seedlings with number of abnormal cells				
			0–6	7–12	13–18	19–24	25–30
19.1	42(21)	2,144	9	6	3	2	1

Data histogramming was performed on the interval with equal length. The number of interval is $M = \left[\sqrt{N}\right] + 1$, where N – is the number of sample data

Table 8.4 Approximation efficiency for the lognormal (LN) and sum Poisson and geometric (PG) models performed by the T-criterion

Dose rate irradiation, cGy/h	Approximation efficiency	
	T_{PG}	T_{LN}
Young		
0	2.64	5.90
0.3	3.97	6.85
1.2	1.99	6.30
19.1	0.06	1.30
Old		
0	6.00	2.90
0.3	2.90	3.50
1.2	11.7	8.60
19.1	0.04	0.16
Young + heat		
0	10.2	6.20
0.3	15.7	13.2
1.2	0.00	0.01
19.1	0.07	0.06

We can conclude that the model "G + P" can be used for the description of seeds on the number of cells with abnormalities. Later, we will use distribution of seeds on the number of cells with abnormalities and analyze the structure of experimental distributions on the model "G + P".

8.2.3 Approximations of Distributions of Pea Cells on the Number of Chromosomal Abnormalities

The above-mentioned methods and some criteria for the assessment of fitting efficiency were used for approximations of the appearance of chromosomal abnormalities in meristem cells. That is as follows: χ^2-criterion (Feller 1957), the AIC criterion (Akaike 1974); the BIC criterion (Schwarz 1978), T-criterion (Geraskin and Sarapultsev 1993). Figure 8.2 presents some of the most effective approximations for young seeds.

8.3 Sample of the Calculation of the Synergic Coefficient

The synergic coefficient was calculated as the ratio of the sum of probabilities of the appearance of an abnormal cell (or chromosome) for each factor separate from the combined one.

8.3 Sample of the Calculation of the Synergic Coefficient

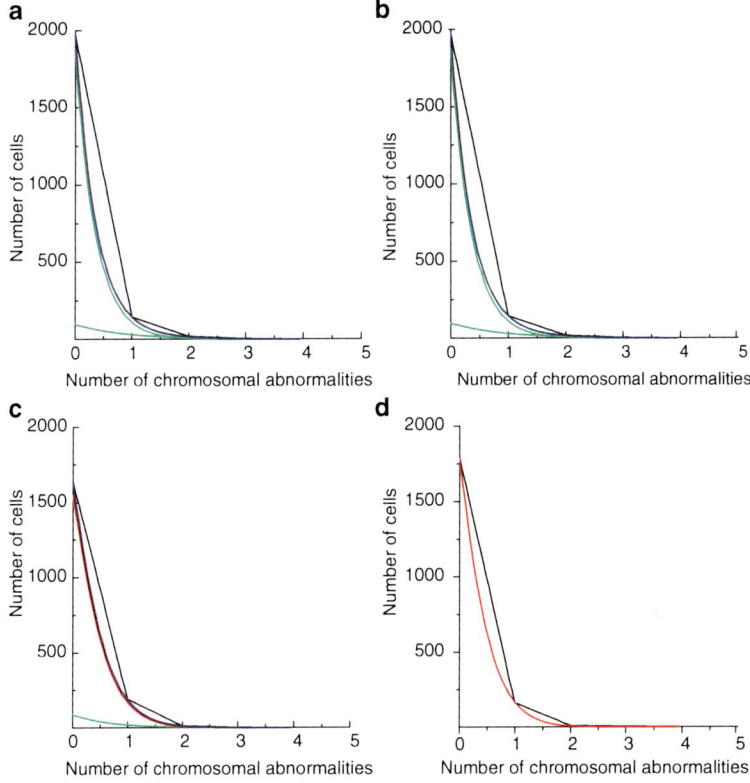

Fig. 8.2 Distributions of meristem cells on the number of chromosomal abnormalities. Group of young seeds. Control (**a**); Irradiation at dose 7 cGy/h: dose rate 0.3 (**b**), 1.2 (**c**), and 19.1 (**d**) cGy/h. Experimental data: *the black line*; model curves: G1, G2 – *green*, P – *red*, and G1 + G2 or G1 + P – *blue lines*

Radiation and heat stresses: Synergic coefficient for survival (see Table 4.2):

$K_{(0.3 cGy/h + heat)} = (1-S)_{heat+0.3\ cGy/h} / \left((1-S)_{young,\ 0.3\ cGy/h} + (1-S)_{heat,\ control} \right)$

$N_{syn}(0.3 cGy/h + heat) = 32.5$

$N_{0.3 cGy/h} = 16.0$

$N_{heat\ (control)} = 7.5$, then

$K_{(0.3 cGy/h + heat)} = 32.5 / (16.0 + 7.5) = 1.38 \pm 0.4;$

$K_{(1.2 cGy/h + heat)} = 62.5 / (7.3 + 7.5) = 4.22 \pm 0.4;$

$K_{(19.1 cGy/h + heat)} = 50.0 / (5.5 + 7.5) = 3.85 \pm 0.4.$

Synergic coefficient for the number of cells with abnormalities (see Table 4.2) as well as for the number of chromosomal abnormalities in cells (see Table 4.7):

$$K_{(0.3cGy/h+heat)} = 8.2/(7.8 + 11.2) = 0.43 \pm 0.01;$$
$$K_{(1.2cGy/h+heat)} = 6.1/(11.4 + 11.2) = 0.27 \pm 0.02;$$
$$K_{(19.1cGy/h+heat)} = 15.4/(9.1 + 11.2) = 0.76 \pm 0.07.$$

Aging effect: Survival $(1 - S)$ (see Table 4.2)

$$K_{(0.3cGy/h+aging)} = 0.75 \pm 0.4;$$
$$K_{(1.2cGy/h+aging)} = 0.97 \pm 0.4;$$
$$K_{(19.1cGy/h+aging)} = 1.04 \pm 0.3.$$

Frequency of cells with chromosomal abnormalities (see Table 4.2)

$$K_{(0.3cGy/h+aging)} = 0.60 \pm 0.08;$$
$$K_{(1.2cGy/h+aging)} = 0.51 \pm 0.05;$$
$$K_{(19.1cGy/h+aging)} = 0.56 \pm 0.03.$$

8.4 Approximation of Distributions for Plantain Seeds Growing near the Nuclear Power Plant

8.4.1 Approximations of the Appearance of Cells with Abnormalities in Meristem of Plantain Seedlings

Here, the approximations of distributions are presented for plantain seeds collected in 1999 on the number of cells with abnormalities in meristem of rootlet of seedlings (see Chap. 5). For the approximations, some of the above-mentioned methods were used. The fitting was performed by means of the FUMILI program.

It is of interest to compare Fig. 8.3 P2 and P3, which correspond to the neighboring sites with different antioxidant status values. One can see that Fig. 8.3 P2 differs from Fig. 8.3 P3 by the decreased distribution values. It can be explained by the small antioxidant status value of plants growing on this site. The higher Poisson sample mean corresponds to the distribution of seeds collected nearest to the nuclear power plant site (Fig. 8.3 P5).

8.4 Approximation of Distributions for Plantain Seeds Growing near the Nuclear...

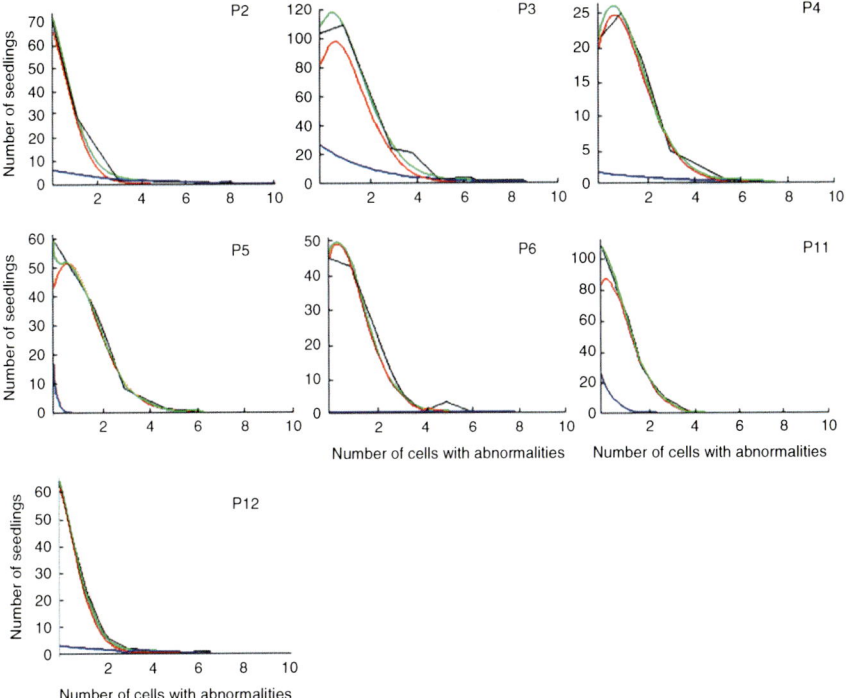

Fig. 8.3 Approximations of experimental distributions of plantain seeds on the number of cells with abnormalities. Seeds were collected in 1999. Plots: experimental data – *black*; Poisson – *red*; geometric – *blue*; sum of Poisson and geometric – *green*. X-axis –number of cells with abnormalities; Y-axis – number of seedlings

8.4.2 Approximations of the Appearance of Proliferated Cells in Meristem of Plantain Seedlings

The approximation of distributions of proliferated cells was performed by the non-linear least squares method. Some criteria were used for estimation of the fitting efficiency. All approximations are satisfied to the χ^2-criterion. The values of the BIC and AIC criteria are given in Table 8.5.

Table 8.6 shows that the approximations for the populations P5, P6, P7, and P12 growing near the radiation sources contain the G-component. We can assume that the radiation factor provided by the NPP and JINR facilities induces instabilities and selection that results in the geometric component of the distribution of proliferated cells. In most cases, the approximations for seeds collected in tested sites have three-components.

The parameters for the best approximations are presented in Table 8.7. The distributions on the number of proliferated cells are shown in Figs. 8.4 and 8.5.

Table 8.5 Efficiency of the approximations of the appearance of proliferated cells (1999)

Populations	3P AIC	3P BIC	3N AIC	3N BIC	2LN AIC	2LN BIC	3LN AIC	3LN BIC	2LN+G AIC	2LN+G BIC
P2	3.26	3.55	3.44	3.64	3.44	3.64	3.03	3.32	3.56	3.81
P3	5.69	5.96	5.48	5.66	4.96	5.23	4.73	5.00	4.10	4.43
P4	4.07	4.32	2.80	2.95	2.80	2.95	2.53	2.75	2.7	2.9
P5	7.43	7.68	5.18	5.36	5.18	5.36	4.52	4.74	4.54	4.83
P6	5.72	6.02	5.03	5.23	5.03	5.23	4.59	4.89	4.46	4.7
P11	3.70	3.83	5.63	5.74	5.63	5.74	4.74	4.87	3.32	3.47

All models are in agreement with the criterion χ^2 ($p > 0.05$)

Table 8.6 The best approximations and their efficiency for the modelling of the appearance of proliferated cells in plantain meristem

Population	1998 Model	1998 Efficiency	1999 Model	1999 Efficiency	
P1	3LN	AIC = 3.16			
		BIC = 3.25			
P2	2LN	AIC = 4.06	3LN	AIC = 3.02	
		BIC = 4.13		BIC = 3.64	
P3			G + 2LN	AIC = 4.73	
				BIC = 5.00	
P4	G + LN	AIC = 3.46	3LN	AIC = 2.53	
		BIC = 3.50		BIC = 2.75	
P5			G + 2LN	AIC = 4.56	NPP
				BIC = 4.74	
P6			G + 2LN	AIC = 4.59	NPP
				BIC = 4.89	
P7	G + 2LN	AIC = 3.62			NPP
		BIC = 3.75			
P8	2LN	AIC = 3.04			
		BIC = 3.09			
P9	2LN	AIC = 3.04			
		BIC = 3.09			
P10	3LN	AIC = 2.54			Chemical
		BIC = 2.58			pollutions
P11	3LN	AIC = 1.62	G + 2LN	AIC = 3.32	Chernobyl
		BIC = 1.48		BIC = 3.47	track
P12	G + 2LN	AIC = 3.72	G + 2LN	AIC = 3.52	JINR
		BIC = 3.85		BIC = 3.05	facilities

The sources of mutagenic pollutions are given

8.4 Approximation of Distributions for Plantain Seeds Growing near the Nuclear... 161

Table 8.7 Parameters of the model which describes the appearance of the proliferated cells in meristem according to lognormal law

Site	Number of seeds	Number of ana-telophases	χ^2/df	m(LN1)	m(LN2)	m(LN3)	N_{LN1}	N_{LN2}	N_{LN3}
1998									
P1	153	726	7.00/8	7.0	12.0	23.0	0.69	0.13	0.06
P2	167	942	1.12/6	5.3	–	22.0	0.61	0.00	0.04
P3	152	518	0.72/7	2.5	12.0	–	0.48	0.32	0.00
P7	156	763	0.45/6	1.0	10.0	24.0	0.29	0.41	0.17
P8	149	1,047	0.28/6	2.0	10.4	–	0.30	0.38	0.00
P9	148	528	0.25/4	2.5	10.1	–	0.14	0.21	0.00
P10	167	231	2.50/5	3.0	9.50	22.5	0.21	0.42	0.08
P11	153	342	0.63/4	4.0	12.0	23.0	0.32	0.13	0.03
P12	148	1,805	1.99/5	12.0	22.0	33.0	0.29	0.19	0.40
1999									
P2	500	2,228	1.67/7	8.0	22.5	33.0	0.12	0.14	7.7e-3
P3	500	3,827	2.21/7	13.5	28.5	39.5	0.24	0.37	0.06
P4	500	1,035	0.45/5	12.0	22.5	33.0	0.11	0.03	0.02
P5	500	2,209	4.21/3	12.0	22.0	33.0	0.13	0.04	0.15
P6	500	2,385	4.50/11	13.3	23.0	37.0	0.21	0.02	0.06
P11	500	2,220	1.14/4	9.0	18.0	3.0	0.33	0.15	0.08
P12	200	832	3.31/4	10.0	22.0	3.9	0.39	0.04	0.19

The comparison of the expected probability with the observed values was provided on the criterion χ^2

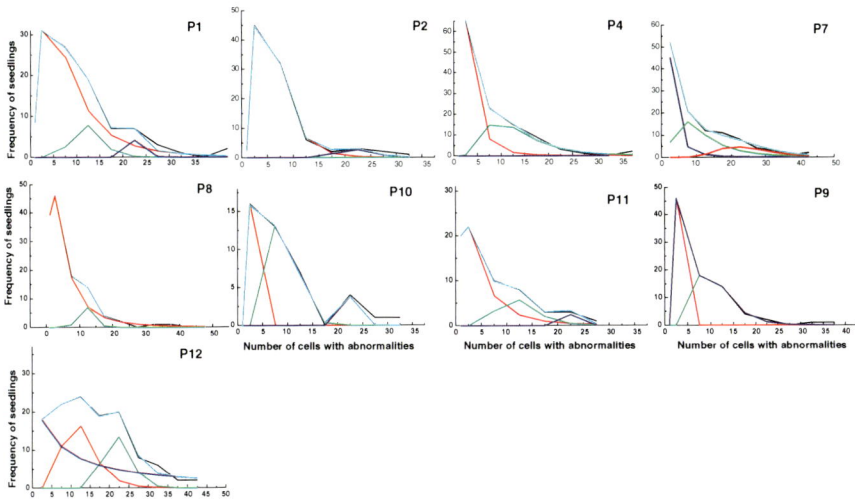

Fig. 8.4 The plantain seeds experimental distributions and the best modelling ones on the number of proliferated cells (1998). Approximations were performed using criteria χ^2, AIC, BIC, T. Seeds were collected in populations: P1, P2, P4 – P12. Plots: experimental data – *the black line*; modelling (geometric or lognormal) – *the red, green*, and *blue lines*; sum of the modelling plots – *the light blue line*. X-axis – number of cells with abnormalities; Y-axis – frequency of seedlings

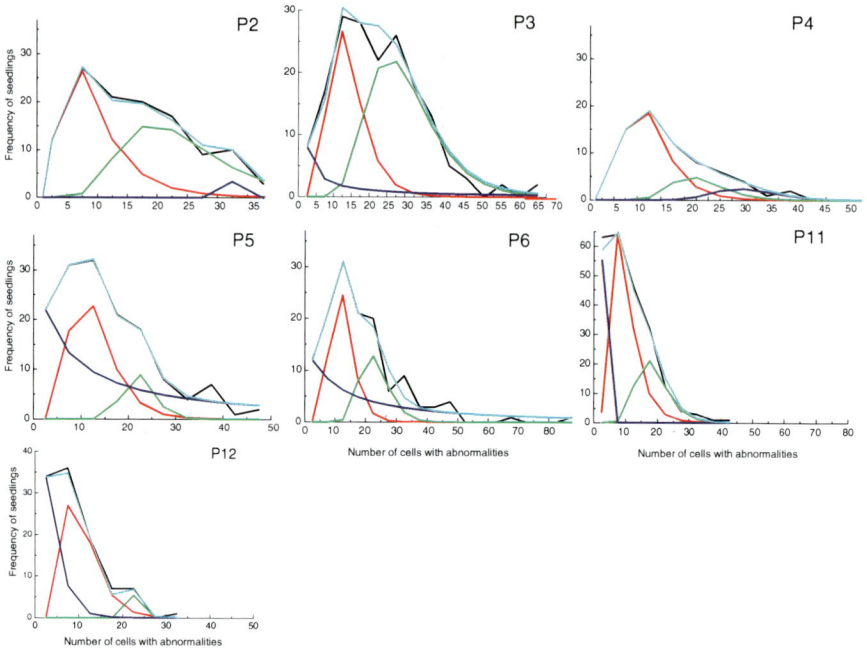

Fig. 8.5 The plantain seeds experimental distributions and the best modelling ones on the number of proliferated cells (1999). Approximations were performed using criteria χ^2, AIC, BIC, T. Seeds were collected in populations: P2 – P6, P11, and P12. Plots: experimental data – *black*; modelling (geometric or lognormal) – *red, green, blue*; sum of the modelling plots – *light blue*. X-axis – number of cells with abnormalities; Y-axis – number of seedlings

8.5 Analysis of the Correlations of Seed Mortality (1 − S) and Parameters of the Plantain Seed Distributions on the Numbers of Proliferated and Abnormal Cells with the Contamination of Chemical Pollutions in Soil (1998, 1999)

It is known that chemical pollution induces geometric distribution on the number of mutation events (Bochkov et al. 1972). Can they imitate the dependence of the seed distributions parameters on the nuclear power plant average daily fallout?

Table 8.8 shows the calculated doses of the daily external radiation exposure provided by the Balakovo nuclear power plant (The Ministry of Atomic Power of the RF et al. 1998). Here, the data for the populations studied in 1999 are considered because at that time strong dependence of modelling parameters on the nuclear power plant fallout was observed.

Determination of chemical pollution (Table 8.9) was provided by the Joint Institute for Nuclear Research (Korogodina et al. 2000). The method is as follows:

8.5 Analysis of the Correlations of Seed Mortality (1 − S) and Parameters...

Table 8.8 Calculated doses (P) of the daily external radiation exposure provided by the Balakovo nuclear power plant fallout (for populations studied in 1999)

Population	Coordinates X(m)	Y(m)	Isotopes Kr	Xe	I	Dose average daily, P	Dose for the summer months, P
P2, P3	−15,400	−13,200	5.80×10^{-9}	4×10^{-10}	5.6×10^{-9}	1.2×10^{-8}	1.4×10^{-6}
P4	−4,300	−7,700	3.80×10^{-8}	1.6×10^{-9}	3.9×10^{-8}	7.8×10^{-8}	9.5×10^{-6}
P5	280	280	1.6×10^{-5}	2.0×10^{-6}	2.8×10^{-5}	4.7×10^{-5}	0.0057
P6	4,950	4,950	9.4×10^{-8}	2.7×10^{-9}	9.3×10^{-8}	1.9×10^{-7}	2.3×10^{-5}

Table 8.9 Contamination of chemical elements in soil samples

Element	Samples P2	P3	P4	P5	P6	P7	P8	P9	P10	P11	P1	P12
	Contamination, mg/kg											
S, %	9.0	<1.0	7.0	**13.8**	1.0		**11.0**			2.5	2.0	10
K, %	1.0	0.1	0.2	0.4	1.4		1.8			0.8	1.6	1.3
Fe, %	2.4		1.4			2.5	**3.9**	2.5	2.2	3.4	1.8	
	1.9	1.2	2.9	1.8	3.3		**3.5**			2.6	3.4	1.2
Ca, %	11.0		5.7			3.8	–	1.8	**9.5**	4.5	–	
	2.3	1.5	4.2	5.3	2.8		2.7			3.5	1.2	2.4
Na, %	–		0.9			0.9	1.2	0.7	0.8	0.7	1.3	
	0.7	0.4	0.9	0.5	0.4		0.4			0.5	0.6	0.3
Mg, %	2.7		1.0			0.6	1.3	0.8	0.9	1.1	0.5	
	0.7	0.1	1.2	2.1	1.1		1.2			0.6	1.1	0.2
Ti, %	0.4		0.4			0.4	0.6	0.6	0.4	0.5	0.3	
	0.3	–	0.4	–	0.5		0.4			0.2	0.2	0.2
Mn	620		880			620	**1,200**	620	780	560	460	
	290	190	510	590	260		220			100	310	110
Cr	120		**220**			110	**190**	160	150	140	**420**	
	110	30	**160**	70	70		80			70	**170**	–
Zn	**180**		**210**			–	–	–	–	–	50	
	54	<10	**130**	100	20		70			110	40	270
Cu	**50**		–			–	–	–	–	–	–	
	<10	<10	<10	20	**130**		<10			73	**110**	**170**
As	120		270			130	**290**	**200**	**260**	**240**	130	
	<2	<2	<2	40	<2		<2			30	<2	50
Se	<1	<1	3	4	3		20			9	<1	<1
Pb	30		60			–	–	–	–	–	–	
	70	10	20	20	**70**		10			**100**	**80**	20
Br	–		50			–	–	–	–	–	–	
	<1	<1	10	10	30		20			**40**	<1	3
Sr	**220**		160			170	120	110	**280**	**240**	130	
	150	80	**210**	170	**200**		**220**			**320**	180	110
U	10		10			10	10	10	20	10	10	
	5	5	5	10	5		5			10	10	2

(continued)

Table 8.9 (continued)

	Samples											
	P2	P3	P4	P5	P6	P7	P8	P9	P10	P11	P1	P12
Element	Contamination, mg/kg											
Th	5	5	10	10	10		10			10	10	10
			<5				10		5	7		
Hg	<1	<1	<1	<1	<1		<1			<1	<1	<1
Cs	2		4			4	8	3	10	5	3	
	1	0	2	2	0		1			5	3	
Ba	210		310				380	**460**	360	360	400	330
	300	180	380	240	**460**		**400**			270	370	220
Rb	30		40			30	**80**	50	50	50	30	
	69	15	**76**	33	73		**80**			59	80	26
Y	**20**		10			10	**20**	10	10	20	10	
	16	8	23	16	24		9			14	**26**	3
Nb	17		15			18	**39**	20	18	17	18	
	8	4	16	7	8		10			13	14	1
Ce	43		19			45	**103**	62	50	54	21	
La	12		20			**31**	28	–	10	–	–	
I	–		–			–	–	–	28	–	–	
Zr	227	85	295	173	177		83			156	85	50
pH	8.1		7.7			7.6	7.2	7.9	7.9	8.3	7.9	7.40
	8.2	8.3	8.0	7.6	7.6		8.2				7.6	7.5

The top and bottom values correspond to the 1998 and 1999 years, respectively

The tested and standard samples were irradiated immediately by γ-quanta on the microtron MT-25 at the JINR Laboratory of Nuclear Reactions over a period of four hours. Maximum energy of accelerated electrons was 24 Mev. Average stream of accelerated electrons was 15 μA. Distribution of γ-quanta stream on the samples was determined by means of copper monitors. Measurements of the γ-quanta spectra of the irradiated samples were provided by means of semiconductor Ge(Li) – detector with value ∼70 cm^3 and resolution 3.5 keV for line 1,332 keV ^{60}Co and Ge(Li) – detector with value ∼28 cm^3 and resolution 2.5 keV for line 1,332 keV ^{60}Co and thin Ge(Li) – detector with resolution 0.6 keV for line 122 keV ^{57}Co. Accumulation and processing of the γ-quanta spectra were provided on a computer.

The roentgen-fluorescent analysis of the soil samples was provided in the apparatus of roentgen-fluorescent analysis at the JINR Laboratory of Nuclear Reactions. The standard radioisotope sources ^{109}Cd (E = 22.16 keV, $T_{1/2}$ = 453 days) and ^{241}Am (E = 59.6 keV, $T_{1/2}$ = 458 years) were used for the X-rays exciting. Characteristic X-rays were registered by semiconductor Si(Li) detector with resolution 250 eV for line Fe (6.4 keV). For measurement, the method of simultaneous determination of all elements excited by a radioisotope source in saturated layers of matter on uniform calibrate curve was used. The standard samples of soil were used to calibrate plotting.

8.5 Analysis of the Correlations of Seed Mortality (1 − S) and Parameters...

Table 8.10 Significant correlations between distribution parameters and contamination of the chemical elements

	Fe%	Na%	Zn	As	U
mP				−0.95	−0.95
mG					
N_P					
N_G			0.90		
m(LN1)					
m(LN2)		0.90			
m(LN3)					
LN1					
LN2					
LN3					
S%			0.90		
AOS%					

Table 8.11 Partial correlations between the chemical elements contamination in soil sample and distribution parameters

	Ca%	Na%	Mg%	Mn	As	U	Cs	Rb
mG	−0.95		−0.98					
N_P								
N_G		0.98						
m(LN1)								
m(LN2)								0.99
m(LN3)				0.99			0.97	
LN1					−0.99	−0.99		
LN2					1.00	1.00		
LN3								
S%		0.97						
AOS%								

Correlation analysis provided by means of the standard program "Statistics" showed correlation between contamination of some elements in soil samples with some parameters of the seed P- and G-distributions on the numbers of cells with abnormalities as well as LN1, LN2, LN3-distributions on the number of proliferated cells (Table 8.10).

We could conclude that Zn contamination influences the value of the G-subpopulation, but not its sample mean. This analysis showed that no element can imitate experimental dose-dependence on nuclear power plant fallout because it should be on both parameters.

Partial correlations between the chemical elements contamination in soil sample and distribution parameters except for "dose-dependence on nuclear power plant" showed the following picture (Table 8.11). We see that there are no correlations between the two parameters (Table 8.11).

Table 8.12 Correlations between the chemical elements contamination and nuclear power plant daily fallout

pH	0.44	Se	−0.59
Fe%	−0.21	Pb	0.61
Ca%	−0.34	Br	−0.44
Na%	0.31	Sr	−0.13
Mg%	−0.26	U	−0.25
Ti%	0.14	Cs	0.00
Mn	−0.26	Ba	−0.06
Cr	0.25	Rb	0.32
Zn	−0.10	Y	−0.12
Cu	−0.28	Nb	−0.08
As	−0.25		

The hypothesis on the appearance of cells with abnormalities in meristem is based on general biological principles and mathematics. It shows non-linear dependence of distributions' parameters on dose irradiation. Correlation analysis assumes a linear relation between the influence and reaction. It means that we have to look for an analogue of the nuclear power plant daily fallout for tested populations in contamination of chemical elements in soil samples, that is a correlation between dose of nuclear power plant fallout and chemical elements contamination.

Let us look for correlations between contamination of chemical elements in soil sample and dose of nuclear power plant daily fallout $|rC_{el},D*|$. There is no such correlation (Table 8.12).

The main atmospheric chemical pollution has been reflected in the chemical content of the soil, and its influence has not been studied.

8.6 Statistical Modelling of the Occurrence Frequency of Cells with Abnormalities in Blood Lymphocytes of Individuals

8.6.1 Experimental Data on Number of Cells with Abnormalities in Blood Lymphocytes of Samples of Individuals Living in Different Regions of Siberia (Table 8.13)

8.6.2 Examination of the Control Samples of Individuals

The cytogenetic analyses were provided at the medical laboratories of the city of Novosibirsk (1,106 individuals) and in the laboratory of the Institute of Cytology and Genetics of SD RAS, Novosibirsk (40 individuals). The results are presented in Fig. 8.6

8.6 Statistical Modelling of the Occurrence Frequency of Cells with Abnormalities... 167

Table 8.13 Experimental number of cells with abnormalities in blood lymphocytes of the samples of individuals living in the different Siberian regions

Number of cells with abnormalities												
N mph	0	1	2	3	4	5	6	7	8	9	10	11	>12
Novosibirsk (2000–2002): inhabitants; 40 individuals including 5 children and 3 teenagers												
4,000	23	17	0	0	0	0	0	0	0	0	0	0
Samburg (1995–1996): native inhabitant and newcomers Nenets; 113 adult individuals												
10,868	16	32	20	11	7	3	1	0	0	0	4	0	19
Samburg (1997): native inhabitant and newcomers Nenets; 48 adult individuals, 1 child and 1 teenager, the total sample is 50 individuals												
3,579	12	13	11	6	0	0	0	0	0	0	0	0	8
Pour district (1998–2000): native inhabitants- tundra Nenets and Selkoup, 65 adult, 10 children and 5 teenagers, the total sample is 80 individuals												
7,404	14	38	11	4	2	1	0	0	0	2	1	1	6
Pribaikal'e- settl. Maloe Goloustnoe, Listvyanka (2003–2004): Russian inhabitants. Listvyanka: 22 adults, M.Goloustnoe: 64 individuals including 1- teenager. The total sample is 86 individuals												
7,950	5	25	24	16	4	6	0	0	0	0	0	5	5

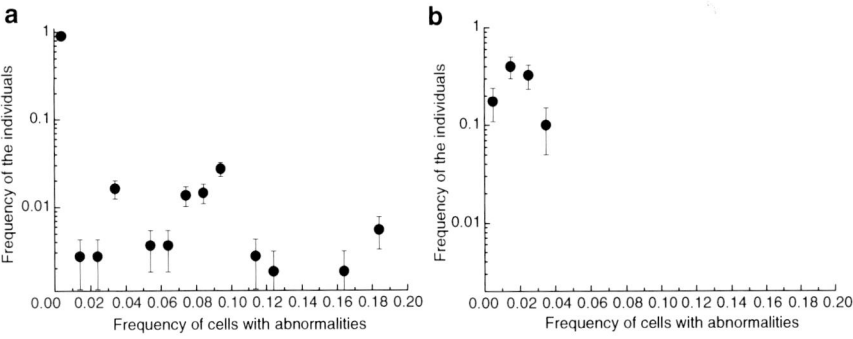

Fig. 8.6 The distributions of the occurrence frequency of individuals (control samples) on the frequency with abnormalities in blood lymphocytes. The data were provided by Novosibirsk medical laboratories (**a**) and laboratory in the Institute of Cytology and Genetics SD RAS ((**b**), our own data). The standard errors are shown

The comparison showed that these distributions are different. Occurrence frequency of the individuals falls dramatically in the interval 0–0.02 of the frequency of cells with abnormalities for medical sample (Fig. 8.6a) whereas it has high values in this interval for scientific sample (Fig. 8.6b).

The medical sample was examined on the analyzed metaphases (Fig. 8.7).

We see that histograms (Fig. 8.7a–d) have dips which reflect omission of abnormalities resulting in the determination of the frequencies. Perhaps the cytogenetic analysis stopped after examination of 12–13 metaphases. Therefore, there are dips after the null class in the medical sample, which leads to significant difference

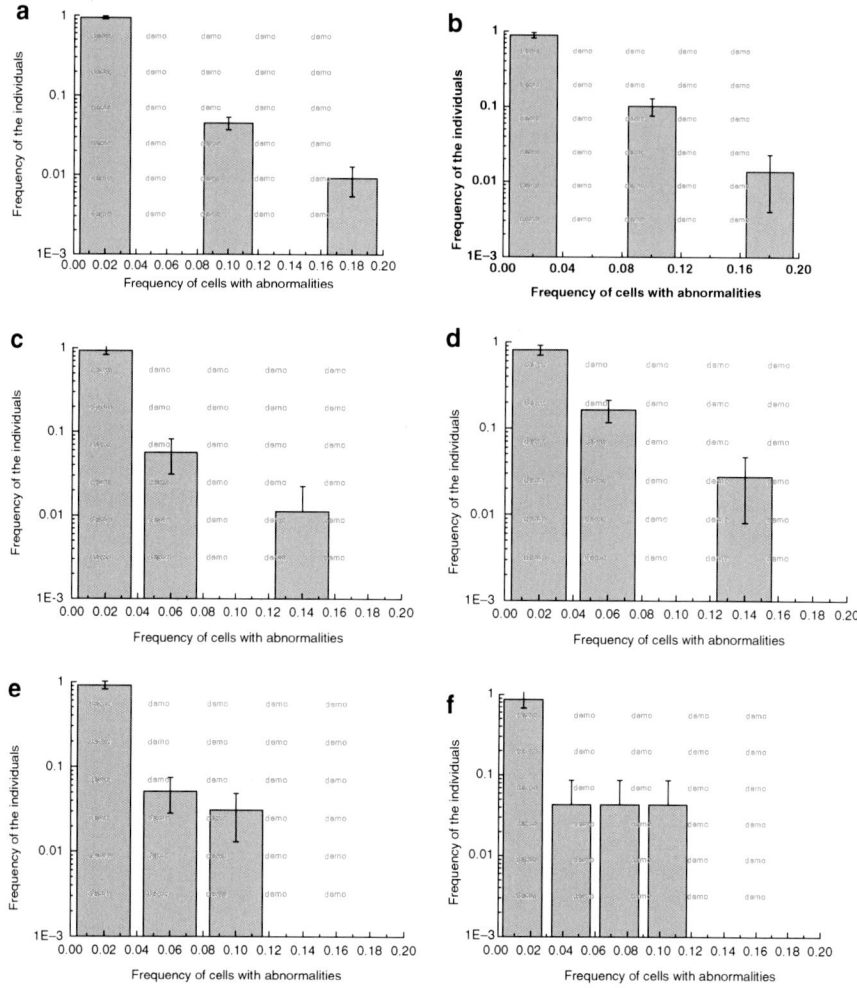

Fig. 8.7 Distributions of individuals on the frequency of cells with abnormalities (medical sample). Number of the analyzed metaphases: 11 (**a**), 12 (**b**), 13 (**c**), 14–20 (**d**), 20–30 (**e**), 30–100 (**f**). The standard errors are shown

between occurrence frequencies of the individuals in interval 0–0.03 frequencies of abnormal cells. In the statistical modelling, the medical data were considered which correspond to the number of analyzed metaphases $N \geq 30$ together with the scientific one where the number of analyzed cells was 100.

8.6 Statistical Modelling of the Occurrence Frequency of Cells with Abnormalities... 169

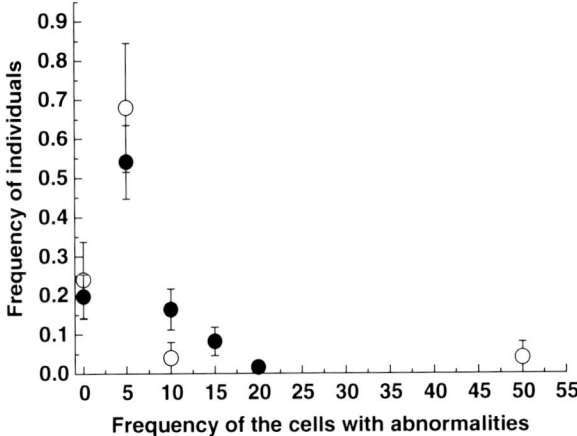

Fig. 8.8 Distributions for the Maloe Goloustnoe and Listvyanka experimental data. For men – *the black circles*, and for women – *the open circles*. The std errors are shown

8.6.3 Examination of the Experimental Data of Individuals Living in the Maloe Goloustnoe and Listvyanka Settlements (Indistinguishability of Distributions for Men and Women)

To analyze heterogeneity of this sample, the distributions for men and women (Fig. 8.8) were studied. The forms of the distributions do not differ.

8.6.4 Approximations of the Experimental Distributions for Samples of Individuals Living in the Settlements of Maloe Goloustnoe and Listvyanka

The methods described above were used for the approximation. The length of the partitioning interval in histograms was varied to the best fit. The model hypotheses P + G, and P were examined. Here, the best results of approximations are presented (Fig. 8.9). The assessment of the approximation efficiency was performed by some criteria including the T-criterion (Geraskin and Sarapultsev 1993) which encourages good efficiency of approximation and fines for using large numbers of parameters. The values of the T-criterion are given in Fig. 8.9 (Maloe Goloustnoe) and Fig. 8.10 (Listvyanka).

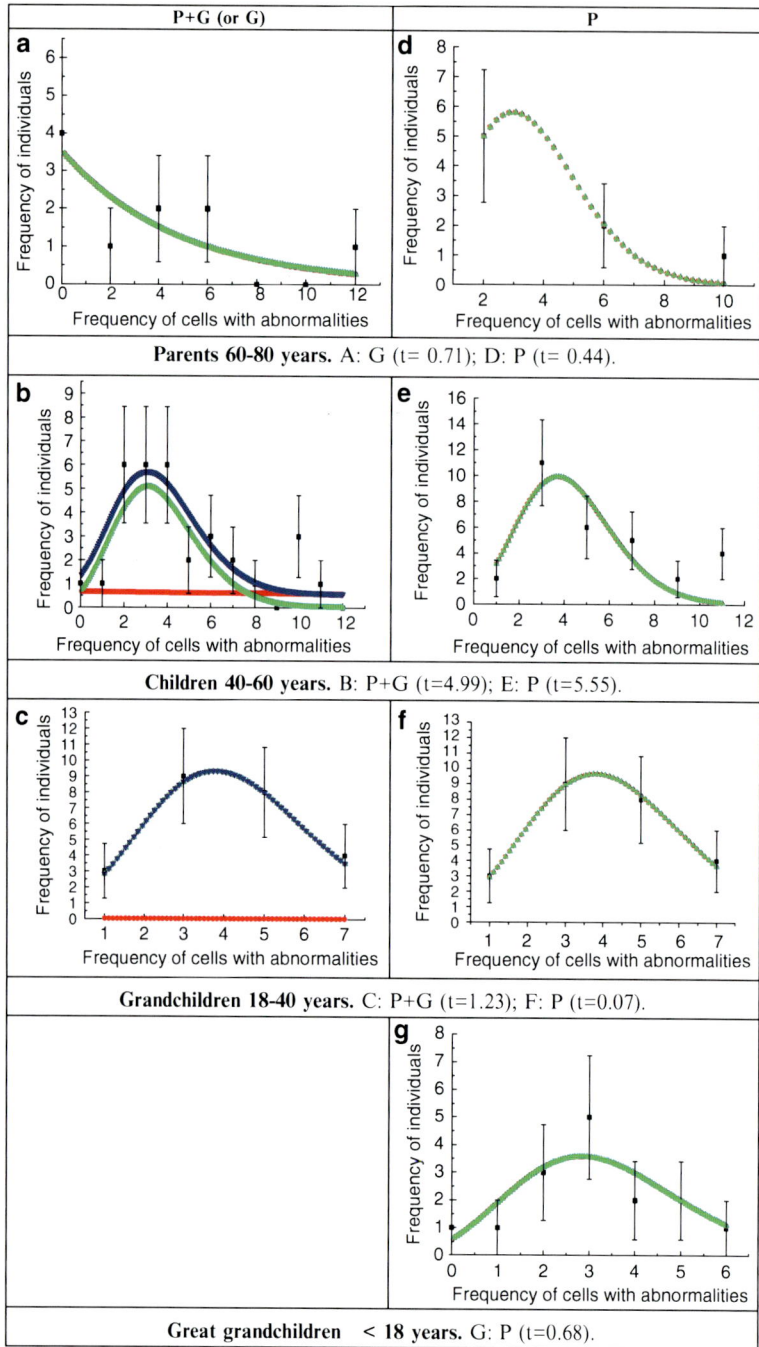

Fig. 8.9 Approximations of the experimental distributions for the samples of individuals living in Maloe Goloustnoe. The best approximations are presented in *bold letters*. The geometric and Poisson components of the model distribution, and their sum are presented by *the red, green* and *blue lines*, respectively (children, G + P model). The std errors for experimental data are shown

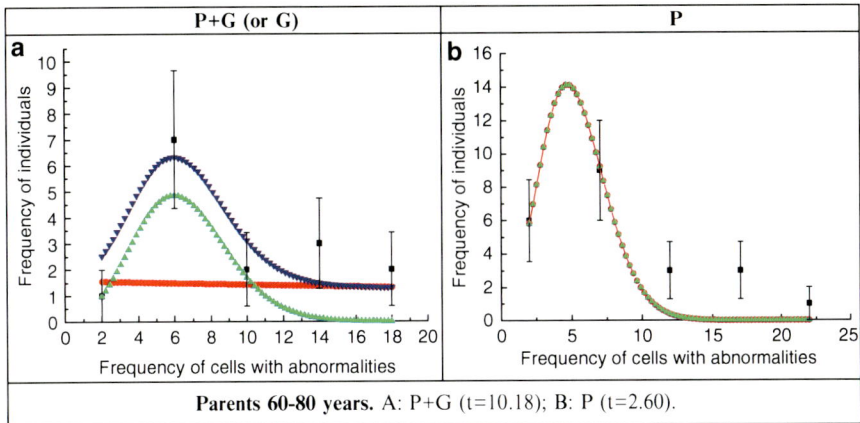

Fig. 8.10 Approximations of the experimental distributions of "parents" samples living in Listvyanka. The best approximations are presented in *bold letters*. The geometric and Poisson components of the model distribution, and their sum are presented by *the red, green* and *blue lines*, respectively (G + P model). The standard errors for experimental data are shown

References

Akaike H (1974) A new look at the statistical model identification. IEEE Trans Autom Control AC-19:716–723

Avriel M (2003) Nonlinear programming: analysis and methods. Dover Publishing, Mineola

Bochkov NP, Yakovenko KN, Chebotarev AN et al (1972) Distribution of the damaged chromosomes on human cells under chemical mutagens effects *in vitro* and *in vivo*. Genetika 8:160–167 (Russian)

Feller W (1957) An introduction to probability theory and its applications. Wiley/Chapman & Hall, Limited, New York/London

Florko BV, Korogodina VL (2007) Analysis of the distribution structure as exemplified by one cytogenetic problem. PEPAN Lett 4:331–338

Florko BV, Osipova LP, Korogodina VL (2009) On some features of forming and analysis of distributions of individuals on the number and frequency of aberrant cells among blood lymphocytes. Math Biol Bioinform 4:52–65 (Russian)

Geraskin SA, Sarapultsev BI (1993) Automatic classification of biological objects by the radiation stability level. Autom Telemech 2:183 (Russian)

Kirkpatrick S, Gelatt CD Jr, Vecchi MP (1983) Optimization by simulated annealing. Science 220:671–680

Korogodina VL, Bamblevsky V, Grishina I et al (2000) Antioxidant status of seeds collected in the plantain *Plantago major* populations growing near the Balakovo nuclear power plant and chemical enterprises. Radiats Biol Radioecol 40:334–338 (Russian)

Korogodina VL, Florko BV (2007) Evolution processes in populations of plantain, growing around the radiation sources: changes in plant genotypes resulting from bystander effects and chromosomal instability. In: Mothersill C, Seymour C, Mosse IB (eds) A challenge for the future. Springer, Dordrecht, pp 155–170

Pytjev YP, Shishmarev IA (1983) The course of probability theory and mathematical statistics for physicists. Moscow State University, Moscow (Russian)

Rakhlin A, Panchenko D, Mukherjee S (2005) Risk bounds for mixture density estimation. ESAIM 9:220

Schwarz G (1978) Estimating the dimension of a model. Ann Stat 6:461–464
Sokolov SN, Silin IN (1962) Determination of the coordinates of the minima of functionals by the linearization method. JINR, Dubna
The Ministry of Atomic Power of the RF, Rosenergoatom concern, Balakovo NPP (1998) The general information on Balakovo NPP. Balakovo NPP, Balakovo (Russian)
Tsyplakov A (2011) An introduction to state space modeling. Quantile 2:1–24 (Russian)
Turchak LI (1987) The basics of numerical methods. Nauka, Moscow (Russian)

Glossary

Anaphase the stage of mitosis and meiosis in which the chromosomes move to opposite ends of the nuclear spindle.

Binomial distribution the **binomial distribution** is the discrete probability distribution of the number of successes in a sequence of n independent yes/no experiments, each of which yields success with probability p.

Bystander effect the **Radiation-Induced Bystander Effect** is the phenomenon in which unirradiated cells exhibit irradiated effects as a result of signals received from nearby irradiated cells.

Geometric distribution the **geometric distribution** is either of two discrete probability distributions:

- The probability distribution of the number X of Bernoulli trials needed to get one success, supported on the set $\{1, 2, 3, \ldots\}$
- The probability distribution of the number $Y = X - 1$ of failures before the first success, supported on the set $\{0, 1, 2, 3, \ldots\}$

Heterogeneity criterion **heterogeneity** and its opposite, **homogeneity**, arise in describing the properties of a dataset, or several datasets. They relate to the validity of the often convenient assumption that the statistical properties of any one part of an overall dataset are the same as any other part.

Homogeneity criterion **homogeneity** and its opposite, **heterogeneity**, arise in describing the properties of a dataset, or several datasets. They relate to the validity of the often convenient assumption that the statistical properties of any one part of an overall dataset are the same as any other part.

Lognormal distribution the **log-normal distribution** is a probability distribution of a random variable whose logarithm is normally distributed.

Maximum-likelihood method **maximum-likelihood estimation** is a method of estimating the parameters of a statistical model. When applied to a data set and given a statistical model, maximum-likelihood estimation provides estimates for the model's parameters.

Meristem cells A **meristem** is the tissue in most plants consisting of undifferentiated cells (**meristem cells**), found in zones of the plant where growth can take place. The cells of the apical meristems divide rapidly and are considered to be indeterminate, in that they do not possess any defined end fate. In that sense, the meristem cells are frequently compared to the stem cells in animals that have an analogous behavior and function.

Normal distribution the **normal** (or **Gaussian**) **distribution** is a continuous probability distribution that is often used as a first approximation to describe real-valued random variables that tend to cluster around a single mean value.

Poisson distribution the **Poisson distribution** (or **Poisson law of small numbers**) is a discrete probability distribution that expresses the probability of a number of events occurring in a fixed period of time if these events occur with a known average rate and independently of the time since the last event.

Proliferation cell growth, division (reproduction)

Sample mean Given a random sample $\mathbf{x}_1, \ldots, \mathbf{x}_N$ from an n-dimensional random variable $\mathbf{X}$ (i.e., realizations of N independent random vectors with the same distribution as $\mathbf{X}$), the sample mean is $\bar{x} = \frac{1}{N} \sum_{k=1}^{N} x_k$

Telophase the final stage of mitosis or meiosis during which the chromosomes of daughter cells are grouped in new nuclei.

Queues theory the **queuing theory** is the mathematical study of waiting lines, or queues. The theory enables mathematical analysis of several related processes, including arriving at the (back of the) queue, waiting in the queue (essentially a storage process), and being served at the front of the queue.

Name Index

A
Abil'dinova, G.Z., 110, 136
Abramov, V.I., 2, 4, 6, 7, 15, 16, 21, 60, 81, 84, 92, 105
Aghajanyan, A., 12, 16, 17
Akaike, H., 152, 156, 171
Akhmadieva, A.K., 12, 22
Alenitskaja, S.I., 87, 104
Alevra, A.V., 88, 106
Alexakhin, R.M., 6, 17, 84, 104
Aliyakparova, L.M., 11, 22
Amiro, B.D., 87, 104
Anisimov, V.N., 11, 22
Antonova, E., 51, 55, 125, 126, 136
Arkhipov, A., 7, 19, 84, 105
Artyukhov, V.G., 84, 104
Arutyunian, R., 10, 18, 26, 55
Atayan, R.R., 60, 81
Auerbach, C., 11, 17
Averbeck, D., 7, 10, 16, 17, 86, 104, 126, 136, 142, 148
Avriel, M., 153, 171
Aypar, U., 10–12, 17

B
Bailey, S.M., 9, 17
Bamblevsky, V.P., 86, 87, 89, 91, 98, 104, 105, 162, 171
Barber, R.C., 11, 12, 17
Barton, L., 66, 81
Batygin, N.F., 141
Bauer, G., 10, 20
Baulch, J.E., 2, 3, 10–12, 16, 17, 19, 22, 145, 149
Bekmanov, B.O., 110, 137

Beljaev, V.A., 88, 104
Belov, A.G., 88, 105
Belyakov, O.V., 2, 4, 9, 17, 21
Beraud, P., 8, 19
Bezdrobna, L., 109, 136
Bezlepkin, V.G., 12, 18
Bird, A., 12, 18
Bliznik, K.M., 8, 11, 16–18, 137
Bochkov, N.P., 26, 28–30, 55, 109, 117, 119, 131, 134–137, 162, 171
Boei, J.J., 8, 17, 61, 63, 67, 81
Boev, V.M., 26, 28, 56, 148, 149
Bogomazova, A.N., 109, 139
Boilley, D., 129, 137
Bolegenova, N.K., 110, 137
Boltneva, L.I., 113, 137
Born, P.J., 95, 104
Brenner, D.J., 5, 13, 15–17
Bridges, B.A., 41, 101, 104
Brown, J., 11, 16, 18
Bryant, P.E., 12, 19
Buchnev, V.N., 87, 104
Bulah, O.E., 87, 104
Buonanno, M., 11, 21
Burdon, R.H., 63, 67, 81
Burkart, W., 9, 21
Burlakova, E.B., 7, 15, 17, 86, 90, 104
Butler, R.N., 82
Buzzard, K.A., 62, 67, 81
Bystrak, D., 27, 56

C
Cao, S., 121, 137
Capocchi, A., 62, 66, 71, 74, 82, 126, 138

Chebotarev, A.N., 29, 30, 54, 55, 134, 137, 162, 171
Cherezhanova, L.V., 6, 17
Chernyago, B.P., 114, 120, 126, 137, 138
Cigna, A.A., 21, 105
Coates, P.J., 7, 9, 10, 15, 19, 22
Condit, R., 27, 55
Corbet, A.S., 27, 55
Crompton, N.E., 9, 21
Cui, J., 7, 19

D

Daillant, O., 129, 137
Davies, K.J., 95, 104
Davis, W.Jr., 66, 71, 81, 126, 137
de Toledo, S.M., 11, 21
Delbrük, M., 12, 21, 149
Deng, Z., 121, 137
Deshpande, A., 9, 17
Dineva, S.B., 60, 81
Djansugurova, L.B., 110, 137
Dodina, L., 86, 102, 104, 146, 148
Dubrova, Y.E., 2, 3, 7, 11, 12, 16, 18, 19, 84, 105, 108, 109, 137
Durante, M., 21, 105

E

Eigen, M., 101, 104
Eliseeva, I.M., 108, 137
Ellegren, H., 7, 18
Emerit, I., 10, 18
Eremeeva, M.N., 109, 138
Eremina, V.R., 117, 137
Etienne, R.S., 26, 27, 56
Evseeva, T.I., 60, 81

F

Feinendegen, L.E., 7, 18
Feller, W., 34, 35, 38, 42–44, 53–55, 74–76, 81, 133, 137, 156, 171
Fesenko, S.V., 84, 104
Filippov, G.S., 8, 11, 20
Fisher, R.A., 24, 25, 27, 55, 147, 148
Florko, B.V., 3, 15, 17–19, 30, 40–43, 51, 53–56, 70, 78–82, 86, 94, 96, 99–101, 104, 105, 122, 132, 133, 136, 137, 144, 148, 149, 153, 154, 171
Folkard, M., 2, 4, 9, 10, 17, 21
Fomenko, L.A., 12, 18
Foyer, C.H., 86, 104

Frankie La, J.V., 27, 55
Frolen, H., 11, 19

G

Gainer, T.A., 121, 126, 138
Galleschi, L., 62, 66, 71, 74, 82, 126, 138
Galloway, A.M., 7, 19
Ganicheva, I., 39, 56, 58, 69, 80, 82, 142, 149
Geard, C.R., 9, 21
Geissler, P.H., 27, 56
Gelatt, C.D. Jr., 153, 171
Gerasimova, E., 66, 82
Geras'kin, S.A., 7, 18, 60, 75, 76, 81, 84, 104, 152, 156, 169
Gerzabek, M., 129, 137
Giaccia, A.J., 62, 67, 81
Gileva, E.A., 26, 28, 56, 148, 149
Gill, V., 63, 67, 81
Gillespie, J.H., 25, 55
Glazko, V.I., 84, 103, 104
Glotov, N.V., 71, 81
Gnedenko, B.V., 31, 55
Goh, K., 9, 18
Goloschapov, A.N., 7, 15, 17, 90, 104
Goodwin, E.H., 9, 17
Gorbunova, N.V., 7, 15, 17, 90, 104
Gossett, D.R., 74, 82, 91, 95, 105
Grandjean, V., 12, 18
Gray, J.S., 26, 27, 56
Grinikh, L.I., 84, 105
Grishina, I., 86, 87, 89, 105, 162, 171
Grodzinsky, D.M., 84, 104
Gubin, A.T., 88, 104
Gudkov, I.N., 7, 8, 18, 59, 60, 62, 63, 67, 78, 81
Gusev, N.G., 88, 104

H

Hall, E.J., 9, 17
Hanawalt, P.S., 66, 82
Harris, R.S., 7, 19
Harris, T.E., 33, 39, 55
Hatch, T., 11, 12, 17
Hickenbotham, P., 11, 12
Hickman, A., 9, 18
Hill, M.A., 10, 20
Hosker, R.P., 88, 104
Houten van, B., 95, 104
Howell, P.C., 121, 138
Hubbell, S.P., 27, 55
Hughes, B.D., 42, 56
Huo, L., 10, 20

Name Index

I
Iida, S., 2, 4, 110, 138
Imanaka, T., 104, 105, 136–138
Iofa, E.L., 108, 137
Ivanov, V.A., 113, 137
Ivanov, V.I., 88, 104
Izrael, Y.A., 113, 137

J
Jaenisch, R., 12, 18
Janssen, Y.M., 95, 104
Jaramillo, R., 9, 18
Jeffreys, A., 109, 137
Jirtle, R.L., 12, 18

K
Kabakova, N.M., 7, 20
Kadhim, M.A., 10, 18, 72, 82
Kalaev, V.N., 84, 104
Kal'chenko, V.A., 6, 7, 21, 84, 105
Kampen van, N.G., 33, 55
Kapultsevich, Y.G., 8, 13, 16–18, 135, 137
Kashino, G., 10, 18
Katosova, L.D., 109, 137
Kaup, S., 12, 82
Kershengolts, B., 90, 91, 103, 106
Khovanov, N.V., 71, 81
Khvedynich, O.A., 84, 104
Khvostunov, I.K., 14, 15, 17, 20
Kilbey BJ., 11, 17
Killender, M., 62, 67, 81
Kim, J.K., 7, 18, 60, 70, 82
Kimura, M., 25, 55
Kirkpatrick, S., 153, 171
Klauder, J., 31, 55
Kolmogorov, A.N., 26, 40, 54, 55, 96, 105, 148
Kolomiets, K.D., 84, 104
Korogodin, V.I., 3, 6, 8, 11, 16–18, 62, 77, 80, 82, 93, 105, 131, 135, 137, 142, 146, 148, 149
Korogodina, V.L., 3, 15, 17–19, 30, 39–41, 43, 47, 51, 53–56, 58, 62, 69, 70, 77–82, 86, 87, 89, 91, 93, 94, 96, 98–102, 104, 105, 130, 132, 134, 137, 142, 144, 146, 148, 149, 153, 154, 162, 171
Korshunov, L.G., 114, 120, 126, 137
Kosarev, E.L., 41, 44, 45, 56
Koshel, N.M., 7, 22
Koutzenogii, K.P., 119, 121, 124, 138
Kovalchuk, I., 7, 19
Kovalchuk, O., 2, 3, 7, 9, 10, 12, 16, 19, 21, 84, 105, 145, 149

Kravets, E.A., 84, 104
Kresson, E., 104
Kuleshov, N.P., 110, 136
Kutlakhmedov, Y.A., 84, 104
Kuzin, A.M., 7, 19
Kuzmina, N., 12, 16, 17

L
Lacassagne, A., 8, 19
Lazjuk, G., 109, 137
Lea, D.E., 13, 19, 28, 56
Lechner, J., 9, 18
Leonard, B.E., 14, 15, 17, 19
Li, M., 11, 21
Liang, X., 7, 19
Liden, K., 116, 137
Lindgren, G., 7, 18
Little, B., 10, 20
Little, J.B., 9, 10, 13, 15, 17, 19, 20
Little, M.P., 7, 19
Liu, Z., 9, 19
Longerich, S., 7, 19
Lorimore, S.A., 9, 10, 15, 19, 22, 31, 42, 56
Lozovskaya, E.L., 89, 105
Luchnik, N.V., 6, 8, 15, 16, 19, 35, 40, 56, 58, 59, 62, 63, 65, 67, 82, 91, 96, 105, 142, 143, 149
Luckey, T.D., 7, 16, 19
Luke, G.A., 12, 19
Luning, K.G., 11, 19
Lyng, F.M., 10, 19
Lyszov, V.N., 88, 104

M
MacCarrone, M., 63, 82
MacDonald, D.A., 10, 22
Maguire, P., 10, 19
Malikova, I.N., 113, 138
Marino, S.A., 9, 17
Martus, P., 26, 55
Maslov, O.D., 88, 105
Matveeva, V.G., 117, 137
Matzke, M., 88, 106
Mazurik, V.K., 7, 17
McClean, B., 10, 19
McClintock, B., 11, 19
McGill, B.J., 26, 27, 56
McNeill, F.E., 9, 19
Medvedev, V.I., 113, 114, 120, 126, 137, 138
Mekhova, L.V., 7, 22
Mellman, W.J., 121, 138
Michael, B.D., 9, 21

Mikhailov, V.F., 7, 17
Mikhalevich, L.S., 109, 137
Millhollon, E.P., 74, 82, 91, 95, 105
Moore, S.R., 10, 18
Moorhead, P.S., 121, 138
Morgan, W.F., 2, 3, 9–12, 16, 17, 19, 20, 34, 56, 145, 149
Morozov, I.I., 7, 20
Moskalev, A.A., 78, 82
Mosse, I.B., 56, 82, 105, 149, 171
Mothersill, C.E., 2, 3, 7, 9, 10, 15–17, 19–21, 62, 63, 69, 70, 72, 82, 142, 145, 147, 149
Motomura, I., 26, 28, 56, 148, 149
Moustacchi, E., 7, 16, 21
Mukherjee, R., 12, 18
Mukherjee, S., 152, 172

N
Nadson, G.A., 8, 11, 20
Nagar, S., 9, 12, 20
Nagasawa, H., 9, 10, 20
Nakamura, A., 9, 21
Natarajan, A.T., 8, 17, 61, 63, 67, 81
Navashin, M., 66, 82
Nepomnyaschih, A.I., 113, 120, 138
Nesterenko, A.V., 2, 4, 84, 106, 108, 109, 112, 139
Nesterenko, V.B., 2, 4, 84, 106, 108, 109, 112, 139
Nesterov, V.N., 109, 137
Neubauer, S., 26, 55
Nevstrueva, M.A., 116, 138
Nikjoo, H., 14, 15, 17, 20
Nikolaev, D., 109, 137
Nilsson, A., 11, 19
Nizhnikov, A.I., 116, 138
Noctor, G., 86, 104

O
Oganesian, N., 10, 18
Ohba, K., 60, 82
Okabe, A., 10, 20
Okada, M., 10, 20
O'Reilly, S., 72, 82
Orr, H.A., 24, 25, 53, 56, 78, 80, 82, 147, 149
Oudalova, A.A., 7, 18

P
Pakulo, A.G., 87, 106
Panchenko, D., 152, 172
Panteleeva, A., 39, 56, 58, 69, 80, 82, 142, 149
Pantiukhina, A.G., 60, 70, 77, 82
Parry, J.M., 7, 22
Parsons, W.B., 8, 16, 20
Pease, G.L., 8, 16, 20
Pechkurenkov, V.L., 2, 4, 7, 15, 16, 21, 92, 105
Perez, M.L., 62, 82, 145, 149
Petin, V.G., 7, 13, 18, 20, 60, 70, 77, 82
Pfeiffer, P., 9, 20
Pfenning, T., 9, 15, 19
Pilinskaya, M.A., 109, 138
Pinzino, C., 62, 66, 71, 74, 82, 126, 138
Plumb, M., 11, 16, 18
Plyusnina, E.N., 78, 82
Ponomareva, A.V., 110–112, 116–118, 129, 138
Portess, D.I., 10, 20
Poryadkova, N.A., 6, 15, 21
Posukh, O.L., 110, 111, 119, 121, 124, 129, 138
Potetnia, O.I., 109, 138
Pozolotina, V.N., 6, 20
Preobrazhenskaya, E.I., 6, 15, 21, 62, 82, 84, 89, 91, 105, 146, 149
Preston, F.W., 26–28, 40, 56, 148, 149
Primmer, C.R., 7, 18
Prise, K.M., 2, 4, 9, 10, 16, 18, 21
Pytjev, Y.P., 152, 172

R
Raabe, O.G., 11, 22
Rainwater, D.T., 74, 82, 91, 95, 105
Rakhlin, A., 152, 172
Ramzaev, P.V., 116, 138
Randers-Pehrson, G., 9, 21, 22
Reed, W.J., 42, 56
Rice-Evans, C., 63, 67, 81
Riches, A.C., 12, 19
Rigaud, O., 7, 16, 21
Robbins, C.S., 27, 56
Rokitko, P.V., 84, 105
Romanova, O., 109, 136
Romanovskaya, V.A., 84, 105
Ronai, Z., 66, 71, 81, 126, 137
Rozanova, O.M., 12, 22

S
Sablina, O.V., 117, 137
Sachs, K.D., 13, 15–17
Sapezhinskii, I.I., 89, 105
Sarapultsev, B.I., 75, 76, 81, 152, 156, 169, 171

Name Index

Satow, Y., 109, 137
Savko, A.D., 84, 104
Sawant, S.G., 9, 21
Schettino, G., 9, 21
Scherbov, B.L., 111–113, 116, 118, 119, 121, 138
Schoen, M., 8, 19
Schuster, P., 101, 104
Schwarz, G., 152, 156, 172
Seabright, M.A., 121, 138
Sedelnikova, O.A., 9, 21
Sen'kova, N.A., 121, 126, 136, 138
Sevan'kaev, A.V., 109, 138
Seymour, C.B., 2, 3, 7, 9, 10, 15, 16, 20, 21, 62, 63, 69, 70, 82, 142, 147, 149
Seymour, R.J., 10, 20, 147, 149
Shao, C., 10, 21
Shaposhnikov, M.V., 78, 82
Shemetun, A.M., 109, 138
Shevchenko, V.A., 2, 4, 6, 7, 15, 16, 21, 60, 81, 84, 92, 105, 108–110, 138
Shevchenko, V.V., 84, 105
Shishmarev, I.A., 152, 172
Sigg, M., 9, 21
Silin, I.N., 152
Sipyagyna, A., 12, 16, 17
Skinner, M.K., 12, 18
Smith, L.E., 9, 12, 20
Snigiryova, G.P., 108, 109, 138
So, Y.H., 7, 19
Sokolov, I.G., 84, 105
Sokolov, S.N., 152, 172
Sprott, R.L., 82
Stamato, T.D., 62, 82, 145, 149
Starodub, G.Y., 88, 105
Stognij, V., 90, 91, 103, 106
Stoian, E.F., 108, 137
Sudarshan, E., 31, 55
Sumner, H., 9, 18
Suskov, I.I., 109, 138
Suzuki, K., 10, 18
Sviatova, G.S., 110, 136

T
Takeichi, N., 2, 110, 138
Tanaka, K., 2, 110, 138
Tew, K.D., 66, 71, 81, 126, 137
Timofeeff-Ressovskaya, E.A., 6, 21
Timofeeff-Ressovsky, N.W., 5, 6, 9, 12, 13, 15, 21, 101, 106, 141, 143, 149
Tsyaganok, T., 109, 136
Tsyplakov, A., 153, 172

Turchak, L.I., 153, 172
Tyuryukanov, A.N., 6, 21

U
Uchihori, Y., 10, 20
Upton, A.C., 7, 21

V
Vaiserman, A.M., 7, 22
Vasil'eva, G.V., 12, 18
Vasiliev, A.G., 26, 28, 56, 148, 149
Vecchi, M.P., 153, 171
Veldink, G.A., 63, 82
Vermeulen, S., 8, 17, 61, 63, 67, 81
Vorob'eva, M.V., 109, 139
Vorobtsova, I.E., 11, 12, 22, 109, 139
Voronina, T.F., 87, 106

W
Waldren, C.A., 9, 22
Watkins, C.H., 8, 16, 20
Watson, G.E., 10, 22
Werner, H.R., 82
Whittaker, R.H., 27, 28, 56
Wiegel, B., 88, 106
Wiley, L.M., 11, 22
Williams, C.B., 27, 55
Winemiller, K.O., 27, 56
Wright, E.G., 9, 10, 15, 19, 31, 42, 56

Y
Yablokov, A.V., 2, 4, 84, 106, 108, 109, 112, 139
Yakovenko, K.N., 26, 28, 29, 55, 109, 117, 119, 134, 136, 162, 171

Z
Zadelhoff Van, G., 63, 82
Zaichkina, S.I., 12, 22
Zhen, Z., 121, 137
Zhivotovskjy, L.A., 71, 81
Zhizhina, G.P., 7, 17
Zhloba, A.A., 109, 138
Zhou, H., 9, 22
Zhurakovskaya, G.P., 60, 70, 77, 82
Zhuravskaya, A., 90, 91, 103, 106
Zimmer, K.G., 6, 9, 12, 13, 21, 141, 149
Zykova, A.S., 87, 106
Zyuzikov, N.A., 7, 22, 86, 104

Subject Index

A

Adaptation, 1–3, 6–8, 15, 83, 93–96, 101, 102, 132, 133, 141–148
   adaptation processes, 1, 2, 15, 23, 24, 53, 54, 57, 74, 80, 83, 97, 101, 103, 134, 141, 142, 144–148
      inter-and intracellular processes, 30, 78–81, 93–96
         scheme, 78–79, 93–95, 132
   adaptive response, 7, 8, 14–16, 69–70, 80
   radioadaptation, 6–7
Aging, 62–69, 71–75, 77–78, 80, 81, 126–132, 135, 145
Antioxidant status (AOS), 86, 89–93, 98–103, 144–146
Approximations, 74, 97, 122, 124, 125, 127, 129–131, 134, 151–156, 158–162, 169–171

B

Background-level radiation (natural background), 2, 6, 24, 83, 85, 88, 102, 145, 147
Biological communities structure, 24, 26–28
   geometric law, 26, 28–30, 33, 34, 37–41, 43, 46–55, 74, 78–80, 103, 124–126, 134, 142–144, 148
   geometric series of I. Motomura, 26–28, 148
   log-series of R.A. Fisher, 27
   logarithmic law, 26
   lognormal distribution of F.W. Preston, 27–28, 40, 148
   lognormal law, 26, 40, 41, 54, 143, 144, 148
   species abundance distribution, 26–28, 84, 103, 148

Bystander effect, 2, 7–10, 12–16, 31, 34, 70–72, 77–79, 80, 102, 128, 129, 131, 135, 142, 143, 145–147
   signaling, 10, 11, 16, 144

C

Cancer, 10, 11, 112, 113, 119
   oncogenic (cells) transformation, 14
Chemical pollutions, 85, 88–89, 93, 102, 103, 162–166
Chernobyl accident, 6, 7, 84, 108–109
Climate conditions, 87, 145
   northern extreme conditions, 107, 110, 128, 147

D

Distribution
   binomial distribution, 42–46, 48–55
   distribution of individuals on the frequency of abnormalities, 41–55
      bell-shaped distribution, 49, 53, 55
      exponential distribution, 25, 36, 42, 49, 53, 54
      Relay conditions, 44–46, 51
   geometric distribution, 26, 28–30, 33, 37–41, 46–48, 50, 53–55, 71, 72, 74, 95–97, 123–126, 132–136, 142–147
   lognormal distribution, 27–28, 40–42, 49, 50, 54, 96, 97, 144, 148
   Poisson distribution, 13, 26, 29–34, 37–39, 51, 54, 70, 71, 73, 74, 76, 77, 94, 95, 124–126, 129, 132–136, 142–144
   type (of distributions), 37, 44, 46, 51, 53, 54, 124, 126, 128, 129, 135, 144

E

Ecology, 3, 6, 15, 17, 83–103, 142, 147
Effect of youth, 107, 131, 135
Epidemiology, 3, 15, 17, 115–121, 147
Evolution, 5, 53, 101, 143

F

Fragmentation (process), 26, 40, 54, 148

G

Gene expression, 2, 9, 12, 16, 86
  epigenetic regulation, 2, 11, 12
  transgenerational signal, 12, 16

H

History
  bystander effect, 8–9
  low-dose radiation effect, 6, 15, 141
  Timofeeff-Ressovsky school, 1, 5, 6, 58
  transgenerational response, 11
Hit principle, 12, 28
Hormetic effect, 7–8, 78, 81
  hormetic coefficient, 78

I

Instability, 2, 7–12, 15–17, 24, 30, 34, 62, 72, 74, 77, 81, 83, 93, 95, 99, 101–103, 108, 109, 114, 119, 122, 125, 126, 128, 129, 131, 135, 136, 141–148
  chromosomal instability, 3, 8, 15, 74, 101–103, 109, 110, 119, 131–132, 135, 136, 143, 145–147
  genomic instability, 2, 7, 9–12, 15–17, 34, 74, 142, 145
  transgenerational instability, 2, 11–12, 16, 128, 147

K

Kyshtym radiation accident, 6, 7

L

Linear no-threshold concept, 7, 14
Low dose-rate radiation effects, 7, 12, 13, 80, 102
Low-dose irradiation, 2, 3, 7–11, 14–16, 57, 58, 69, 77, 102, 142
Lymphocyte depletion, 125, 132, 135, 147

M

Models
  adaptation model, 3, 6–8, 15, 141–148
  Bystander and Direct model, 13–14, 16
  correlative model of multiplication of the DNA damages, 34–39
    geometric-and Poisson components, 36–39, 53–54, 143, 148
  diffusion model, 14, 17
  evolution model, 25, 53
  Fisher's model, 24–25
    micromutationism, 25
  geometric model of adaptation, 24–25
  Microdose model, 14, 15, 17
  model of appearance of cells with abnormalities, 30–34
    Poisson-and geometric components, 26, 33, 34, 53, 54, 94, 132, 133, 142–143, 148
    termination of the adaptation process, 33, 53, 55
  model of the multiaberrant ctll appearance (Chebotarev), 29, 30, 134
  model of the proliferated cells occurrence, 39–41, 54
    lognormal-and geometric components, 26, 54, 143, 148
  mutational landscape model, 25
  probabilistic model, 12–13, 16

N

Non-linear response, 6, 9, 15, 16, 57–70, 73, 80, 81, 91, 92, 141, 142
Nuclear power plant (station), 7, 85, 145

P

Plant (seeds), 57–62, 89
  meristem (of rootlet apex), 60–62,
Process
  microevolution process, 5, 15, 101
  mutation process
    cascade mutagenesis, 8, 11
    mitotic disorders, 8
  radiation-induced processes, 3, 12, 14, 16, 57, 71, 128, 146, 148
Proliferation, 7, 8
  cell cycle delay, 63, 67
  mitotic index, 63, 65
  proliferative (mitotic) activity, 8, 58–60, 62, 84, 90, 143
  resting cells, 8, 39, 40, 54, 62–66, 74, 96, 100, 101, 143, 144

Subject Index

stimulation of proliferation, 7, 8, 16, 39, 41, 54, 58, 63, 65–67, 68, 71–74, 77, 78, 80, 96, 148

Q
Queuing theory, 28, 38

R
Radiation effect (general regularities), 6–16, 80, 145–147
  Balakovo nuclear power plant, 102, 103, 145
  North Siberia, 111–113, 117–119, 135, 136, 147
  Pribaikal'e region, 114, 121, 145
Radiation intensity (dose rate), 7, 9, 80, 81, 143, 144
Radioactive contamination, 6, 28, 145–147
  caused by Chernobyl accident, 84, 108, 109
  fallout of nuclear tests, 109, 113–120
  fallout of operating nuclear plant, 85–88
  operation of accelerator facility, 88
  operation of radiochemical industry, 134–136
Reactive oxidative species (ROS), 62, 86, 126, 129, 145
Repair mechanisms, 7, 15, 34–37, 53, 54, 65, 78, 80, 94, 144
Risks
  accumulation of abnormalities/selection, 102, 131, 147
  risks of instability, 15, 17, 102, 103, 131–132, 147

S
Selection (Darwinian), 2, 24, 25, 32–34, 41, 53, 80, 102, 135, 148
Semipalatinsk nuclear tests, 108–110, 113–114, 120, 121, 128, 129, 135
Sensitivity/resistance, 15–17, 34, 53, 54, 72–74, 81, 102, 103, 133
  humidity, 60, 90, 103, 146
Site characteristics
  30-km zone near the Balakovo nuclear plant, 85–89
  Siberian Far North, 110, 111, 115–117
  South Baikal zone (Pribaikal'e), 113–114, 120
Statistical modeling, 2, 15, 23–55, 70–81, 95–103, 122–136, 142–148
  efficiency of approximations, 154–156
  method of approximations, 151–153
    approximation by the χ^2 minimization, 152
    least-squares method, 152
    maximum-likelihood method, 152
    random search method, 152, 153
Stress
  combination of radiation and heat stresses, 63–81, 91–93, 97–103, 146
  oxidative stress, 63, 71
  radiation stress, 30, 39, 54
  stress factor, 2, 5, 7
Synergic effect, 68, 70, 77, 80, 81, 103, 146
  synergic coefficients, 77, 81, 156–158

T
Target theory, 9, 11, 13, 28
Totskij nuclear explosion, 26, 28

Printed by Publishers' Graphics LLC